Hello.

Published by 404 Ink Ltd
www.404Ink.com
@404Ink

ISBN: 9781912489183
ISSN: 2399-1577

404 Ink co-founders & publishers: Heather McDaid & Laura Jones
Editor and Project Officer: Imogen Stirling
Cover design: Letty Wilson
Inside design: Laura Jones
Interviews for 404 Ink conducted by Imogen Stirling, Heather McDaid and Katie Sharp

Printed in the UK by Stephens & George

Imogen Stirling

A note from the editor

Welcome back.
It's been a while.
Make yourself comfortable.

Issue 5 is ready and waiting for you.

Maybe you've been here before or maybe you're new to 404 Ink. Either way, we're glad to have you with us. A lot of changes have been afoot, and we want to share them with you. If you remember, we took a break after releasing Issue 4 to step back and consider the direction of the magazine. How can we shake things up? What can we do differently?

Well, here you have it.

To begin with, you'll see we've downsized. No more doorstop-sized magazines arriving in your post. Instead we've opted for a thinner, slicker and more colourful design. We've increased our number of feature pieces alongside our fantastic showcase of new writing in order to fully explore the chosen theme – or should I say, themes. You see, we've opted to start publishing our magazines in themed pairs, starting with the first duo of Space and Earth, the latter being our upcoming Issue 6 later this year. There seems no grander pairing, with the scope of the two topics really letting the imagination run riot – as you'll see within this issue's pages. To top it all off, we have the amazing Letty Wilson creating the beautiful cover artwork for both issues.

Oh, and then there's me.

Hello! Hi! I'm Imogen. I'm the new Project Officer here at 404 Ink. I've been on board since the start of the year, and what a joy it's been. I've had such a warm welcome into the 404 Ink world which seems to consist of doing one cool thing after another (it's not every day that you get to interview an astronaut, after all – but more on that later).

Together, we've made what I must say is a belter of a magazine. As always, we present an array of poetry, fiction and non-fiction. We received more submissions for this issue than ever before, making the selection process incredibly tough, but the range of pieces we've ended up with is stellar, with their interpretations of space being diverse, clever, poignant and funny. We've got space hipsters and cyberpunks. We meet Stephen Hawking and Jeff Bezos (competing in an intergalactic song contest and doing... questionably). We're permitted an insight into the world of living with chronic pain and witness poetic despair over the ways we use space to create borders and enforce isolation. Our writers document cocktails created to mark the end of the world, the best (and worst) writing utensils to use in space, and a certain doomed sex mission of two lovestruck geckos. There are poems to make you wonder and stories to make you laugh.

Meanwhile, this issue sees us include our largest number of features to date to really dig into the broad and fascinating topic of space. And we really went to town here. We spotlight two brilliant space-themed books – *Oor Big Braw Cosmos* by John C. Brown and Rab Wilson and *Space is Cool as Fuck* by Kate Howells – and chat with their authors about inspirations and anecdotes. We chat with Khadija Osman, head honcho at new inclusive bookshop Round Table Books, to discuss the importance of creating inclusive spaces within publishing. We interview REAL LIFE SPACE FLIGHT PIONEER Brian Binnie, record-setting winner of the Ansari X Prize about his journey above Earth, and INCREDIBLE NASA LEGEND Adriana Ocampo, who is Science Program Manager – that includes leading the Juno mission to Jupiter and New Horizons mission to Pluto. No biggie.

We've often kept the 404 Ink magazine as a separate strand of the publisher, but that seemed silly in retrospect, so we've introduced a 404 Ink corner where we catch up on what our authors have been doing and keep up to date with all things 404. Here, we interview Chris McQueer about the BBC adaptation of *Hings*; introduce you to Nadine Aisha Jassat and her debut poetry collection, *Let Me Tell You This*; dive into Elle Nash's raw and perceptive debut *Animals Eat Each Other*; and catch up with Michael Lee Richardson of the Queer Words Project, six months on from the publication of *We Were Always Here*.

As well as bringing 404 to the magazine, we've revamped our Patreon with yours truly at the helm. Back to regular updates, behind the scenes looks and more, it's the perfect place to not only subscribe to outstanding new writing twice a year, but also keep in touch with the process, the people, and the oddities of running a publisher. Never want to miss a future copy? You can subscribe over on our Patreon: patreon.com/404ink. We hope to see you there.

I'm so chuffed to be part of the 404 Ink team. After watching and admiring their work from a distance, it's a real pleasure to be involved. Putting together this magazine has been inspiring; my head now crammed full of people and stories and facts. I guarantee you'll be left with new-found wonder at our incredible universe.

So, pour a drink (we have the recipes), sit down and lose yourself in stories of space.
We hope you enjoy the read.

Three.
Two.
One.

Blast off.

Imogen x

Imogen Stirling
is a spoken word artist and theatre maker.
Her 5 * show #Hypocrisy debuted at the
Edinburgh Fringe 2018 before transferring to
London and is touring around the UK and internationally. She works as a freelance proofreader
and editor, and translates French film scripts.
@imogen_stirling / imogenstirling.com

NON-FICTION

Dos and Don'ts of Writing in Space: A History

Tom Pickles

 DO

understand how pens work on Earth, and why they might not in space

Fountain: internal reservoir of India ink is fed onto nib through a capillary action. Prone to leak if used at high altitudes due to changes in air pressure.

Ballpoint: quick-drying ink is fed onto a ball, which rolls over the page as you write. This action requires gravity to work properly, which is why you can't use one upside down (for very long, at least).

Marker: 'felt' tip is fed from the reservoir by a wick, prone to drying out. Viscosity of ink is subject to vary in a pressurised environment, and so viewed as unreliable.

 **DON'T**

get swept up by the hype of the Fisher Space Pen

Few stories of the Space Race are as neat as the idea of the American space programme spending a fortune developing a pen to write in zero gravity, while their Soviet counterparts were using a pencil. Sadly, it's not true; though a pen was developed by an American to write in space, it was not a project endorsed by NASA. Inventor Paul Fisher funded the development himself, then sold the product (both the 'Space' pen, and the higher spec AG-7 model) to both NASA and the Soviet programme.

 DO

admire the excellence and versatility of the Space Pen

Having said all of the above, the Fisher Space Pen does do everything it claims, and has found many uses in extreme environments on Earth. Essentially a ballpoint pen, the Fisher Space (and AG-7, etc) Pen uses a pressurised cartridge and gel-like ink – the result being a pen that works in zero gravity, upside down, and at temperatures from -45°C to 120°C.

DON'T

underestimate the waste in using a normal, wood-cased pencil

If you tried to sharpen a pencil in a zero g environment, not only would the shavings float away but you would also need to wedge yourself somewhere in order to get the purchase required and not just spin yourself around.

DO

remember that graphite is both flammable and a conductor of electricity

Following the catastrophic fire in the Command Module of Apollo 1 during a launchpad test, NASA greatly increased restrictions of flammable materials in crewed areas of the rocket.

DON'T

forget about public perception and 'value'

Prior to the launch of Gemini 3 in 1965, it was reported that pencils to be carried on the flight had cost $128.90 each (the budget showed that the cost been $4,382.50 to purchase 34 of the pencils). Though this cost was largely due to development of a housing that could be used with a space suit, the headline figure was poorly received, and even challenged in Congress.

DON'T

believe the hype about a pencil being preferred as the 'Russian Space Pen'

Though early cosmonauts carried grease pencils, they were quick to adopt the Fisher Space Pen once available.

 DO

see if you can catch a glimpse of the so-called 'Pencil Nebula'

Properly known as NGC 2736, the Pencil Nebula was discovered in 1835, and is 815 light-years away.

✖ DON'T

take the previous list of pen problems as definitive (...sorry...)

Felt-tip pens were used on the Gemini, Mercury, and Apollo missions, as well as on the Space Shuttle. However, they were noted to work poorly in the low-pressure environment on the Skylab space station and were not relied upon.

Standard ballpoint pens have been used on Russian/ESA missions aboard Soyuz rockets, as astronaut Pedro Duque notes, apparently with no problems.

Chris Hadfield has also noted on Twitter that he used mechanical pencils, space pens, and markers on the International Space Station, which also has a printer and label maker.

Despite the fame of the Fisher Space Pen, it is contended that the most used writing tool on the Apollo missions was the unheralded Garland mechanical pencil.

✖ DON'T

forget that a pen can be more than a writing tool

After inadvertently snapping off a circuit breaker switch in the Lunar Module, Buzz Aldrin used his Duro felt-tip pen to activate the control and fire the engine to leave the Moon.

✔ Finally, DO

take a moment to marvel that there is no record of any poetry having been written in space before 2009, when International Space Station Flight Engineer Koichi Wakata added his entry to a chain poem (No. 25, Third Series, Space Poem Chain). Sadly, though this appears to have been written by hand, I don't know with what (unless he has replied to my tweet since publication, in which case, I do know, and you don't).

Tom Pickles is from Edinburgh. His previous writing can be found in the procedures guides for one of the UK's largest life insurance providers.

The Space Between The Sheets

Rhiannon Walsh

Two months ago, as I moved my bed away from the paper-thin wall connecting my room to the bedroom of the teenage boy who lives next door, I thought: This. Is. It. This new layout will positively impact absolutely everything else in my life. I'll be closer to the window so I can wake up naturally with the sun. My bed will be further away from the plug socket which means no more meaningless 1am Instagram scrolling. Not to mention, with all that extra floor space I've gained, I'll be able to start my day with yoga, a green smoothie in one hand and an acai bowl in the other.

Fast forward to now and, of course, none of those positive habits (or, god forbid, #ProductivityHacks) have materialised. Just like when I first moved in and promised myself to adhere to a strict No Electronics In The Bedroom rule, depicted by a sad wicker basket placed in the hall to hold any laptops or phones that even *thought* about bringing their electromagnetic frequencies and Wi-Fi waves into the threshold of my sacred space of Zen. I eventually caved after three days because I really needed to check Jeff Goldblum's star sign as I lay staring at the ceiling (he's a Libra, in case you're wondering).

At the time of rearranging my furniture and taking a deep, hard look at my burgeoning internet addiction, I was obsessed with wellness YouTubers and TED Talkers who rose from their beds at 4:30am to a glass of hot lemon water and an hour and a half to spend working on their passion projects, before they even had to get dressed for work. I gave it my best shot but quickly realised the downsides, the most considerable involving scalding my hand on the kettle as my eyes tried to make sense of being awake four hours ahead of schedule, or fighting existential burnout at 8am because I'd spent 90 minutes worrying about what my passion project actually *is*, or if I even had one at all?

With that, I saw the dream of becoming a wellness influencer sponsored by GymShark leggings and essential oil diffusers wither away, and the possibility of turning my bedroom into a sanctuary of meditation, productivity and minimalist artwork die before my eyes, just like the Swiss cheese plant I'd bought to oxygenate the room, but had forgotten to water.

My bedroom went back to being what it always was, a space to spend at least one third of my day, but more importantly, a room designed specifically to hold the most beloved and important item in my life right now: my bed.

The truth is that I'm tired. A lot. Clinically, chronically tired. Chronic Fatigue Syndrome and Fibromyalgia will do that to you. I've come to a point where I'm too tired to feel guilty about not using my mornings to meal-prep and write 10 Things I'm Grateful For Today. I'm tired of feeling that for some reason I am a person who has less of their shit together because of that. I'm too tired for morning routines that encompass anything more than attempting to alter my appearance resembling Balthazar from Buffy the Vampire Slayer, to that of a more socially acceptable human one. I am certainly too exhausted to do anything involving passion that isn't passionately hitting snooze on my alarm clock and cursing the fact another day has rudely begun without my permission.

A good day for me is one where I can get out of bed, make some food that isn't pre-packaged tortellini, manage to think clearly for the most part and walk to the shops. A great day is one where I can work in the office without coming home early and an excellent one is when I can do all of this and fit in some socialising come night-time.

A bad day is when I remain confined to the space between my sheets.

My bed feels as though it is the epicentre of my life; it is a place I've longed for, utterly detested and spent more time in than any other location on the planet. Ask any chronic illness sufferer and they will confirm its importance, being somewhere we find ourselves back in, time and time again, regardless of our mood, health or the day's circumstances.

The portal that exists between my soft under sheet and the warm comfort of the duvet and what it represents is one that remains unchanged regardless of flat moves or altered bed positions. It can be replicated in hotels, Airbnbs and friends' couches, but nothing comes close to the original. A safe and familiar space that signifies rest, pain and pleasure, depending on the day.

When I'm in the middle of a Fibro flare-up, in a Cocodamol haze accompanied by low mood and even lower self-esteem, I often think of the space of my bed as one of confinement, where I am held against my will and where time stands still for me as life zooms past the windows. But recently, I've been trying to remind myself that my bed is much more than that.

With every hour of duvet incarceration comes an hour of long, easy conversations with my partner, beneath the sheets, where time stands still in a different way. Or an hour of knowledge gained from exploring late night Wikipedia rabbit holes and Myers-Briggs personality quizzes completed in the blue light reflection of my laptop.

The setting of my bed is one that thrives on extremes.

It is where I've shuddered awake from the most treacherous of dreams and where I've laughed so hard I've snorted at the worst of Netflix comedies. I've lain sobbing in the foetal position from inexplicable pain and lain star-shaped in the centre of my mattress, dead to the world in a coma-like slumber. It's a place where the most visceral of fights have occurred and the best make ups have played out.

My worst traits are portrayed here more than anywhere else, as I frequently fall asleep in toast crumbs I was too lazy to brush off. And it's where Murphy's Law is a common guest as I regularly get my period early just after I've changed the sheets.

The scene is a hotbed of mismatched pillowcases and sheets that don't seem to know what 'fitted' means. Of hot water bottles, orthopaedic pillows, phone cables and fluffy socks I watch myself put on every night but find myself filing missing person reports for the next morning.

It is somewhere that has enveloped me in comfort as I lay hungover, surrounded by the remnants of my bad decisions, or cried broken-hearted from the decisions of others. And a setting that has felt empty, unreal and inconvenient as the never-ending time bomb of insomnia sets in.

My bed is loyal to its inhabitants, remaining unchanged and exactly as I left it throughout the day, awaiting my return to embrace me beneath its fort of white cotton and polyester blend. I've been at my most vulnerable here, wrapped in a towel fresh from the shower as my wet hair drips onto the blankets while I try to garner as much energy as possible to attend pre-made plans, or when I've read up on too many symptoms matching mine on WebMD and sat aghast at all the terrible things that can go wrong with the human body.

It is a space that, in essence, stays the same. The scene only changing due to the mood or condition of its owner. It can tell a thousand stories. The unmade bed represents a rushed morning. The bedside table adorned with a Himalayan salt lamp and SAD-quashing alarm clock depicts good intentions, while a lone vibrator and empty pizza box on the floor portray even better ones. The bed moulds itself to the room, to be whatever it is required to be, from a napping station to a poor trampoline or a work-from-home office space. Ultimately, however, it is a hub of routine and habit, carrying out its pivotal role of providing a space of comfort and protection at the end of each day.

It is not without its magic though. Each night I gaze enthusiastically at the A3-sized vision board made up of Photoshopped pictures of Timothée Chalamet and I attending the Oscars together, an East Village townhouse selling for $2.3 million and huge cut-outs of I AM RICH and I LOVE MY JOB affirmations. Even though my envisioning techniques are always slightly out of focus due to my short-sightedness, this nightly ritual serves its purpose in creating a world in my mind where I am well enough to partake in these fantasies and, despite my conditions, live a life that is worth living. It acts as something I can set my mind to and focus my attention on as I lie in pain, fatigue or crippling brain fog. It is those moments of imagination and the dreams that often follow that allow me to be whisked away from the confines of my bed and into a world of possibility, vitality and enchantment.

Nowadays, as I am forced to look back at my brief stint of wishing to replicate the daily routines of Oprah or motivational influencers, I've come up with some of my own conclusions. While I don't think rising before the sun with a high-intensity workout and goal-mapping session is bad or destructive, for most of us with chronic illness or disabilities it is an unattainable and unrealistic achievement that makes us all the more aware of our energy and mobility levels often being less than those of others. Similarly, the popular rule to keep your bed as a place solely for rest leaves out the people who spend most of their time there. Unless you are going to personally entertain me for the entirety of my time under the covers, I do not wish to hear how I shouldn't create, work or watch television in the space that I am involuntarily restricted to.

What I do agree with within this prominent wellness movement, however, is that our bedrooms are indeed sanctuaries. But it is important to remember the definition of 'sanctuary' is different to us all. For some, it is simply a place to sleep or have sex; for others, it's a spot to watch 14 hours of Grey's Anatomy, surrounded by empty Maltesers packets. For people like me, it can be all of these things, as well as the space we find ourselves in more often than we'd like. Unlike the typical definition, our sanctuaries capture elements of all of life's moments, not just the clear-headed peaceful ones.

My sanctum is less concerned with mantras, wind-chimes and special editions of *The Secret*, and more focused on providing a place of solace, support and reassurance during the day's ups and downs.

The space between the sheets is somewhere I've slept, cried, came, written, laughed, eaten, read, shaken and sobbed. While I can temporarily alter it with a set of fresh, warm sheets from the dryer or a new velvet blanket, it exists primarily as a place to hold me at my most vulnerable and powerful.

Our beds are often more loyal to us than any lover, job or friend could ever be. Versions of them are with us from our first day out of the womb to our moment of departure.

While they may be required more by some, they are private, closed off, personal spaces, cherished and honoured by us all.

Rhiannon Walsh currently works in TV as a researcher and enjoys writing on the side. Her favourite topics are chronic illness, health and pop culture. She is the editor and curator of The Chronic Project zine and currently resides in Glasgow with her rescue house rabbit Duncan. Twitter: @TheNapKween.

Space Cocktails

Lucy Jane Santos

Think of cocktails and, more than likely, imagery of impossibly glamorous people, smoky rooms, and bootleggers will pop into your head. Or perhaps it's something closer to unsavoury bars with lurid coloured abominations masquerading as cocktails.

But these mixed drinks are so much more than that: they can also be used to tell stories of the past. They can be a window into many different types of histories, not least because they are reflections of the intentions of various peoples: the establishment that commissions them, the person that makes them, and even the customer who is meant to drink them.

Sometimes, the name of the cocktail itself can offer us an insight into the most unlikely parts of history. For many cultures, the naming of something gives it power, substance, and meaning and it is no different for cocktails.

Lucy Jane Santos raises a glass to two important dates in space history, decades apart, that led to the creation of two very different cocktails, and looks at some stories of alcohol and space.

The Comet Cocktail (1910)

- Take a snifter of French vermouth, a jagger of applejack (a type of apple-flavoured alcoholic drink which was traditionally freeze distilled) and some cracked ice.
- Mix together in a glass and serve.
- This mixture is guaranteed to lull your fears and allow you to watch the sky with the bravery of a single soldier against an army.

This very simple cocktail recipe was created, rather surprisingly, to mark the end of the world.

1910 was full of excitement for space lovers. The year had started with The Great January Comet which was so impressive that it was even visible in daylight. But the real eagerness was for the long-awaited return of Halley's Comet, which was picked up by telescopes in early April.

The comet's return had long been anticipated; in fact, it had been known to return periodically since the eighteenth century when the English astronomer Edmond Halley had first studied its course and predicted its reappearance. The comet had been observed and recorded by astronomers since at least 240 BC, and clear records had been made by Chinese, Babylonian and medieval European chroniclers but were not recognised as reappearances of the same object at the time. In 1705, Edmond Halley used his friend's, Sir Isaac Newton's, laws of gravity and motion and compiled a list of 24 comet observations to show that those of 1531, 1607 and 1682 were all the same object. Halley, therefore, predicted that the same comet reappeared every 75-76 years. He was proved right, although he didn't live to see it, when the comet reappeared in December 1758.

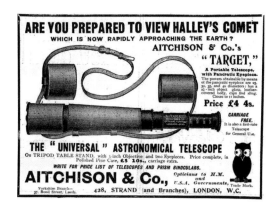

In 1910, the anticipation turned to concern when astronomers revealed that Earth would pass through the comet's 25 million km long tail. This was a worry because in February of that same year the Yerkes Observatory in Wisconsin used the technique of spectroscopy, when light is analysed to show the composition of celestial objects, to discover that one of the things the tail was made out of was cyanogen: a deadly poison which is part of the cyanide family. Cyanogen gas is an irritant to the eyes and respiratory system. Inhalation of the gas can lead to headaches, dizziness, nausea, vomiting, loss of consciousness, convulsions, and death, depending on exposure.

And with a 25-million-kilometre-long tail full of it, that was a lot of predicted exposure.

This genuine, albeit unfounded, concern was stoked by newspapers and lurid publications that featured pictures and descriptions of people choking to death. *The New York Times* reported that the French astronomer Camille Flammarion has said that the cyanogen 'would impregnate the atmosphere and possibly snuff out all life on the planet.'

This growing atmosphere of fear was irresistible ground for charlatans and entrepreneurs who offered, amongst other things, comet insurance, anti-comet pills for just $1 a box, gas masks and even anti-comet umbrellas.

The comet also appeared in advertisements for a diverse range of products such as soap, tea, coffee, yeast, champagne, lightbulbs and chewing tobacco.

And as the 19th of May, the day it was known that the Earth would pass through the comet's tail, came closer both anticipation and fear grew. People prepared for the end of the world: women in Chicago were said to be frantically stuffing the cracks around their windows and doors with paper, a man in South Africa advertised space for one person in his anti-comet shelter, and, tragically, people ended their lives rather than risk being exposed to the poison.

But not everyone reacted in the same fatalistic way to the threat of the end of the world. Some people decided to hold a party.

At the great New York hotels, The Astor, St Regis, Gotham and The Knickerbocker, the smart set could attend special Comet Parties with menus themed to the comet craze. Men and women dressed in their finest clothes and jewels awaited the world to end (even if it was rather tongue in cheek) while eating *clams a la comete* and *omeletes a la comete* and even potatoes shaped like stars.

And many, many more faced the problem with a stiff drink in their hand.

The Comet Cocktail described in contemporary reports as being a 'soothing concoction' was very popular, but alternatives such as the Cyanogen Highball, which was apparently guaranteed to protect the drinker from the deadly gas, were also available.

How this would work in practice wasn't specified, but presumably, from the amount of alcohol in the recipe, the plan was to get too drunk to notice.

At the Claypool Hotel in Indianapolis, a bartender named Charles Swiggett, who went by the moniker 'infuser of libations', created a cocktail of a heavy cinnamon liqueur, both kinds of vermouth, the whites of an egg, and some 'odds and ends that he wouldn't divulge.' It was apparently very popular.

People gathered on the roofs of hotels and houses, many with a Comet Cocktail in their hand, and waited. In the end, the comet appeared, everyone was rather underwhelmed, and they went home. But the Comet Cocktail remains with us.

Comet Lovejoy

In 2014, NASA scientists studied the Comet Lovejoy, which was popularly named 'The New Year's Comet', as it made its way round its elliptical orbit of the solar system. The scientific analysis revealed that the comet had 21 different complex molecules and was emitting vast quantities of ethyl alcohol and glycolaldehyde on its journey – essentially booze and sugar.

In a press release issued by NASA, Nicolas Biver from the Observatoire de Paris in France was quoted as saying: 'We found that Comet Lovejoy was releasing as much alcohol as there was in at least 500 bottles of wine every second during its peak activity.'

Comet Lovejoy, named after Terry Lovejoy, the amateur astronomer that discovered it, appeared in the sky during 2014 and 2015, passing closest to the Sun on the 30th January 2015.

Unfortunately, it isn't expected to return for another 8000 years.

Moonwalk (1969)

- Take 30 ml of grapefruit juice (freshly squeezed if you have it), 30 ml of orange liqueur, three drops of rose water and enough champagne to fill your glass to the brim.
- Combine the grapefruit juice, orange liqueur, and fragrant rose water in an ice-filled cocktail shaker.
- Shake semi-vigorously for thirty seconds and strain into a chilled champagne flute.
- Top the mixture up with the bubbly champagne and enjoy, preferably while staring up at the night sky.

Joe Gilmore was head bartender of the world famous American Bar at The Savoy Hotel, London, when he created the Moonwalk. The American Bar is the oldest surviving cocktail bar in Britain. It opened in 1893 in the original riverside part of the hotel and moved to its current position in 1904. In 1969, Gilmore was at the top of his game. He had started as a trainee there in 1940 and had been swiftly promoted to the highest position at the tender age of 33. He held the role until he retired and was often brought back as a guest bartender for special occasions.

Over 600 million people had watched the crew of Apollo 11 land and walk on the Moon in late July 1969. The mission was the culmination of decades of investment and excitement. The promise of space exploration heavily influenced popular culture at the time. This was seen in television shows like *The Jetsons* and *Star Trek* as well as in architecture and interior design. For Gilmore, it influenced one of his most famous signature cocktails created to both celebrate the mission and the safe return of the crew.

The Moonwalk soon became the most requested drink at The Savoy. A flask of the concoction was even sent on a Pan Am flight to Houston, which was where the returning astronauts were held in quarantine for three weeks. Legend has it, this was the first thing Neil Armstrong, Buzz Aldrin, and Michael Collins drank after they were allowed out of isolation. Even if this isn't quite true, there is a letter, proudly displayed in The Savoy's museum, from Neil Armstrong himself thanking Gilmore for the tipple.

Of course, such an important historical event influenced many more cocktails, but this one is by far the tastiest. And if you visit The Savoy today, you can still go to the American Bar and try the drink. This cocktail, a snip at only £100 a glass, is made of Vintage Grand Marnier, Dom Perignon, a sugar cube, grapefruit bitters and orange flower water, but they probably will make you the original version if you ask nicely.

Alcohol and Space

Buzz Aldrin broke all the rules about not having alcohol in space. NASA were so strict that even products like mouthwash were not allowed if they contained alcohol. Buzz wasn't desperate for a drink, though, he had something very different in mind. The plastic packages that he smuggled in containing bread and wine were because he wanted to perform a communion ceremony during a planned radio blackout. This religious ceremony wasn't broadcast because of fears of complaints, but Buzz was apparently successful in his mission.

Although NASA still will not allow their astronauts to drink in space, there are others – especially some well-known 'astropreneurs' or space entrepreneurs – who have dedicated themselves to solving the problems that microgravity – the scientific term for zero gravity – can cause

alcohol. And as Richard Branson moves (perhaps) closer to launching Virgin Galactic – the world's first commercial space flight – the need for this research increases.

In 2015, Japanese brewing and distilling company Suntory sent samples of six of its whiskies into space on board a shuttle departing for the International Space Station. Their researchers were interested in how microgravity would affect the ageing process of whisky.

Open Space Agency, a collective of astropreneurs, also started working with the whisky company Ballantine's in the same year. Their brief was to design a glass that could be used in zero gravity. Their solution – the Space Glass – was unveiled in September 2015.

Utilising 3D printing techniques and innovative design, the Space Glass has been scientifically proven to work in space. Instead of actual space, the team used the ZARM Drop Tower testing facility in Bremen, Germany, which simulates zero gravity, to show that their design worked.

Ballantine's even created a special blend of their whisky, the 'Special Batch Ballantine's Space Whisky', which is a more concentrated whisky with enhanced spice and honey notes. The enhancement was necessary because space can change your taste buds, making it harder to determine flavours.

Lucy Jane Santos is a freelance writer and historian with a special interest in popular science and the history of everyday life. She writes & talks (a lot) about cocktails and radium. She can be found on Twitter @lucyjanesantos_ Her debut non-fiction *Half Lives* will be published by Icon in July 2020. www.lucyjanesantos.com

POETRY

Disorderly Conduct

Kathryn O'Driscoll

The Moon is molten
with a sunray-thin custard skin
confining it globular as it hangs, ripe.

I'm going to pluck the Moon.

Squeezing satellites until they burst,
running lavacicle vein-routes down my palms
and creating forest fires in my elbow pits;

I won't care. I'm going to crush the Moon.

Plush inside,
rich and overly vanilla
compared to the searing cinnamon of the Sun.

I am going to eat the Moon.

With teeth sunk through cellular veils,
with Saturn's rings dug into bloodied gums
and Plutonian cavities,

I will destroy the Moon.

If I digest a planetary system,
intestinalise stardust,
and choke on the pulsation of a supernova,

if I eat the universe, the world, the Moon –

maybe then
I'll be full.

Kathryn O'Driscoll is a South West based performance poet who performs regularly, including on billings with Shane Koyczan, Joelle Taylor, Danez Smith and Vanessa Kisuule. She's a slam champion and has competed at several UK Poetry Slam national competitions. Her poetry often focuses on disability, mental health, loss and gender-based issues.

Stargazer

Alycia Pirmohamed

I used to think of prayer as surrender:
a white flag folded into a waxwing
cuff – a flight response.

As a child, my father would name
the dark and nebulous knot of evening.
How the bright was first: god

until the burial, when it became the shape
of his father. Two decades later, I am
driving to Banff

at night during a thunderstorm
when a shadowed white-tailed deer
tumbles onto the highway

wet and rickety on her legs.
I look skyward for something: some star,
some god, some streak of light

to clear the fog. Something father enough
to illuminate a staggering deer's slick hide.
As if burnt hydrogen, as if

the *kabristan* in the sky, will fend off
all I fear. As if I am that child still resting
her chin at the windowsill.

Alycia Pirmohamed is the author of the chapbook *Faces that Fled the Wind* (forthcoming, BOAAT Press). She is a 2019 recipient of the 92Y Discovery Poetry Contest, and a winner of the 2018 Ploughshares Emerging Writer's Contest in poetry. Alycia received an MFA from the University of Oregon, and is currently a PhD student at the University of Edinburgh.

Advice to Returning Astronauts

Marcus O'Shea

1.

Periods of deep space isolation
breed contempt from those in distant stations,
towards those who remained back safe on Earth.
This is why the system has experienced breakdowns over time.

When you spend months floating

    weightless

        and almost entirely

            alone

                among an infinite stretch of stars

it is hard to cope with the return to Earth
and the feeling in your withered limbs
that you are being crushed
beneath the now unfamiliar pressure.

It is a normal reaction to feel some resentment
to loved ones who remained behind,

to feel frustrated at the way they move
so easily
without noticing that the Earth below them
drags them down.
To hate the way they breathe
without problems
air filled with foreign bodies,
unpurified and unfamiliar,
while you choke at the taste of it.
To misunderstand
that it is not out of spite they move
so swiftly
in the morning
when unfortunate waves of solar radiation
have left you feeling like your blood
is sore and heavy in your veins.

Some astronauts cannot move past this,
and sign up for second tours.
They make a career
out of the loneliness in the stars.
We should not begrudge them for this choice.
When home has become something alien for you
it is understandable to flee to a place of comfort.

2.

Certain astronauts have described
a side-effect of the conditions.
Where months spent staring
at a sea of infinite darkness,
terrifying in its complexity
and smeared with celestial lights
becomes too much for your mind to process while you're up there.
And instead, when they return to Earth,
at the most inopportune moments,

they

    find

       themselves

          drifting

             from the surface

and floating again, somewhere far above
(although, they would be quick to correct me,
there is no real concept of "above" where they go).
It can be anywhere,
in a shopping mall
with children grown so big while they were away,
parade,
nightclub.
The rush before climax in a darkened room.
The flashing lights
and sounds and voices fade away
and instead they see nothing.

But the constellations
they gave new names to,
while stuck up there in space.

Marcus O'Shea is a writer and food activist. They have performed on three continents, been arrested on two, and once attempted to purchase Winston Churchill's blood at auction. They live in Edinburgh and their work has been shortlisted for the Reader Fiction Prize, the Bristol Short Story Prize and Bath Flash Fiction Prize.

The Space Gecko Project

Stuart Kenny

In 2014, *The Telegraph* published an article headlined: "Five Geckos Freeze to Death on Space Sex Mission."

Since this incident, however, new evidence has come to light.

The Space Gecko Project, a foundation set up to honour the deceased geckos and celebrate their lives, has come into possession of the diary of one of those geckos, which was salvaged from the wreckage of the Foton M4 Satellite on which they travelled.

Its contents, which chart the life of the gecko in question from his birth to his ill-fated, yet crucially romantic final moments in space, prove that this was in fact no "sex mission" because two of the geckos, at least, were very much in love.

What follows is the last entry written by that lovestruck gecko, experts believe, in his final few moments on the ship. The poem was then finished by scientists from the foundation involved in the project based on footage recorded on board the satellite.

It is the climactic entry in the aforementioned gecko's diary, which has now been immortalised in the form of multimedia spoken word show *The Space Gecko Project*, run by the foundation to tell the full story of the star-crossed geckos, and set the record straight.

May its publication bring peace to the gecko who wrote it, to his beloved partner, and capture them in their true light, which global media chose, somewhat callously, to overlook.

Regards,

Stuart Kenny, Grant Robertson, and Lewis Gillies

The Love Letter from the Foton M4 Satellite

I only signed up because you were going.

You put our names down as a joke
but then
before you know it,
we're locked hand-in-hand
on the rocket launch pad,
used to just be friends
now so much more than that.

We took off not knowing that we'd never land.
Rosy cheeks,
stomachs flipping like an acrobat,
you whisper in my ear that even if it doesn't go to plan
that'd be just fine,
cause you're already in your dream land.

See, you were always the gecko next door for me
and I couldn't believe
that the papers could be so naive.
You know The Telegraph described this as a "sex mission"?
Completely missing my admission
that you give me definition.

Love is risky business,
but it feels like we've completed the impossible mission.
Like Tom Cruise!
But without the scientology.
We're just having top fun,
studying astrology.

When I met you
I was just a broken gecko bulb
and then you came along
like a gecko technician
and made me shine again.

You're the ink to my fountain pen
and my only suspicion,
something I think about now and then,
is why a gecko like you would be with a gecko like me?

But you just bat away my inhibitions
with some caramelised crickets.
You know that that's my favourite food,
because it's our Friday night tradition.

But I'm still wishing
I could go back and fix those papers' omission.

Tell them you're like the Lizard of Oz,
the world's best magician!
You've put a spell on my heart,
not even death will do us part,
yeah this is Romeo and Juliet:

Gecko edition.

Before I met you, I felt like 50 Cent in Poundland
walking round the world empty handed.
Not half there,
then you met my stare.
Your fingers fit in mine and your laughter
was
so
contagious.

Fun all over
and such a love for exploration.

Nothing could
contain us,
pain us
and nothing could restrain us,
now I'm up in space, just want to stare at your an...

...gel-like qualities.

And I always told you that I loved you to the moon and back.
That if I could pluck the stars out the sky for you
I'd leave it jet black.
That one day we'd fly to Saturn's rings and never come back,
and then we got in the backseat
of our little gecko Cadillac.

No reptile disfunction here!
No, we can perform,
but this is love, not lust,
and just before
you say:

"I know that we're cold-blooded,
but you make my heart feel warm."

And look - there's Saturn now! You couldn't planet better.
There's the Milky Way! Let's try and eat it together!
My mum always feared I lacked a little endeavour,
she'd be so proud I've found a superstar like you for my forever.

Well...

I guess it's kind of romantic that it's up in space we're gonna freeze.
We've come so far!
I still remember our first kiss - back in your parents' tree.
After I stood outside your window
playing 'Hello' by Lionel Richie,
on a little gecko boombox,
and it filled you up with gecko glee.

You always loved your music.
So when I got down on one knee
I hired four rhinoceros beetles
to play that insect classic, 'Let It Bee'
You told me to shut my eyes and count slowly up to three,
then put your own ring on my finger, and said:

"Let's go explore the galaxy."

Now there's whole worlds outside our window
and you're still looking at me.

But now the ice is creeping over the walls and floors
like a Hollywood scene,
so we're going to take it into the third person,
so we can watch it like the big screen:

Sparkling ice spreads like running water down the street.
Our geckos shuffle closer together
as the ice takes hold of their feet.
Sets them frozen in position,
where their mission's gonna end,
and they link up their hands so tight,
and the warmth is like a godsend.

The ice is wrapping round their knees,
they know they're never getting free.
So they just pretend they're back in their teens
in her parents' silver birch tree.

"Babe," says the guy gecko. "I'll love you forever!"

And she says:

"Of course I love you too,
But call me babe again and ice or not your balls are staying blue!"

And then they both start laughing,
that laugh they've laughed a thousand times,
on gecko dates at gecko fairs where
they once won a gecko prize.

In gecko pubs and gecko clubs,
now gecko trips into the skies,
through gecko lows when times were tough,
and more so during gecko highs.

But now with water in their eyes,
they're running out of time for real goodbyes.

He just can't bring himself to say it.
His little gecko heart just can't take it.
Geckos don't have eyelids,
so at least their gaze will never break,
there's no mistaking.

As the ice takes over,
starts to wrap around their shoulders,
he's just glad that in his little life he ever got to know her.

And she just smiles her gecko smile
and speaks their final truth.
She says:
"I don't mind being frozen solid,
if it means I'll be frozen in time with you."

Hands forever together they kiss,
and the last thing they feel is the warmth of each other's lips
as the ice locks them in time so perfectly,
comets pass and supernovas blast
like flowers blooming fervently,
and Venus sings to signal,
the Gladiator's certainty
that this kiss between two geckos,
will echo
in eternity.

Go to page 76 for a peek behind the scenes of The Space Gecko Project.

Stuart Kenny is a Scottish spoken word artist and creative writer. A regular face on the Edinburgh poetry scene, The Space Gecko Project, complete with music by Grant Robertson and illustrations by Lewis Gillies, debuted at Hidden Door Festival and was described as "possibly the most wholesome event for Scottish poetry in 2018" by The Skinny.

Cyberpunks

Stephen Watt

Fox Mulder's *"I want to believe"* poster
hangs upon her kitchen wall
eclipsed by looming, wuthering clouds;
stencilled pinions smutting their video sun.

She knew that they would come. The cyberpunks.

She stops sweeping her artificial lawn,
adjusting the iBall
to relate scale, the pretext for this visit,
and the truth. Her circuited-tongue
dials on a touch-tooth –

fangs for emergency calls.

Pluto was voted out of the planetary club.
Demoted. Reportedly apoplectic
at their demotion to status of dwarf planet.

This would be their apocalyptic answer.

She watches the Shepard Scale of spacecrafts
stria the skyline, musical notes
suspended on staves
and waits for the symphony to begin.
Missiles, hidden in the bristles of the brush,
she pointedly trains on the sky.

Stephen Watt is Dumbarton FC's Poet in Residence and the current Makar for the Federation of Writers (Scotland). His books include *Spit, Optograms*, and *MCSTAPE* with a fourth collection entitled *Fairy Rock* due out in late 2019. Stephen reviews for punk site Louder Than War and The Wee Review.

hydrogen-hellos in retrograde

Ashley Cline

sometimes i wish that spontaneous human combustion

was a lot less spontaneous and a lot more compassionate
because god, how i'd light up a room—

like, that girl in the corner? don't mind her,
like, that girl sipping oxygen? leave your lighters at home,

like, that girl casually burning? no need for flash photography
because, not to brag or anything,

but god, would it be spectacular—

like a real-life red giant on Earth,
turning telescopes into kaleidoscopes and back again just for fun

like a sympathetic supernova dressed in black at the bar buying everyone shots of the theory
that energy isn't lost, only borrowed, and isn't that comforting?

like knowing that nothing burns much brighter than a hydrogen-hungry star
and, god, isn't that what we're told we all are?

don't we all casually read "star on the rise"
and think, god, *that could be me?*
or is it more—

that could be me, or at least
it could be me if it weren't for this thing,

you know, this thing where i turn my body into a graveyard,
like every backyard in America made into sentry and cemetery

and god, how i can turn rocks into epitaphs like sculptor turned
child, like child turned mile-marker

like, have you noticed how the shadows worship the sun
because they know they can only exist in the negative space of warmth?

and they are okay with that—

and i think i should want to be okay with that, too
i think i should be—

or maybe that's just me,

maybe that's just me orbiting orbiting orbiting
the vacuum of small talk and cold bedsheets

because, god, how we nearly burned up
in our return from the space between our fingertips

because god, how we nearly burned through
each other's tongues,

like a whisky watered down with copper sulfate,
burning green like the river after undressing,

jilted and jealous of the storms that salt the rim of the glass sitting on the nightstand table,
because energy is never lost, only borrowed

and it's not so much that i think about you

but i do think about whether or not you think about me thinking about you,
and i think about whether or not you think about that,

or if you think about how an exploding star gives more than it takes,
like a cosmic baptism, like life after death

like you never knew when to
say *when*

like, that girl in the corner? like, that girl combusting?
like that girl dressed in black and

practicing hydrogen-hellos in retrograde?
like maybe

she was never burning for you.

Ashley Cline is an avid introvert and full-time carbon-based life form currently living in South Jersey. You can find her essays about music and feelings at Sound Bites Media. She graduated from Rowan University in 2013 with a bachelor's degree in Journalism, and her personal best at all-you-can-eat sushi is five rolls in eleven minutes.

Bold

David G Devereux

For me, Star Trek stands for hope
The hope that humanity can go beyond its petty squabbles and
Boldly go where no-one has gone before*

*(I'm mostly going to be talking about The Next Generation because I think philosophically and in terms of production value it's a better show but that's a different argument)

Star Trek shows us what we can be when we grow up
Intrepid space explorers, yes, and
A species that is above greed, above prejudice
A simple creed that all life is valuable and extraordinary
That this universe is filled with philosophies
And it is possible for them all to coexist
And to learn, and benefit from each other**

**(with the exception of the Borg, the Romulans, the Dominion, Klingons most of the time, Species 8472, the Mirror Universe...)

It sounds so simple
No money, no religion
Your achievements are based on your abilities
Not who you are, or where you came from

And yes, you can say this is all a fantasy
But so far Star Trek has predicted mobile phones, video calling, touch screen technology
And NASA now has an elementary tricorder
So perhaps a socialist utopia
Is not so far out of reach

And yes, Star Trek is just a TV show
But the Bible is just a book
And Hamlet just a play
Format is irrelevant
When you have something important to say

For me, Star Trek stands for hope
Sometimes life on this small, blue-green planet can be claustrophobic
Fearing that our troubles will take us down a road to the undiscovered country
From which there is no voyage home
Star Trek looks to the sky, defiant and bold
And says; "that is where we will go"
Where no-one has gone before

David G Devereux is a writer, musician and audio producer based in Glasgow. From 2016-2018 he wrote and produced science fiction audio drama podcast 'Tin Can' and is currently show-runner and producer for 'Middle:Below' – an audio drama podcast about ghosts, cats, and the slightly odd people who look after them.

liminal being

Sean Wai Keung

that night a mouse appeared out of a tiny crack in the wall of my room at first i thought it was a very big spider since it was dark + i had never seen a mouse in my room before but it started shuffling around against the sides of the walls away from the crack + i couldnt stop looking at it + it was only after some time had passed i finally seemed capable of thinking *thats a mouse* but beyond that thought i had no other ideas i mean i have no feelings towards mice no particular fear or love truthfully i had never thought much about mice besides the times i had seen them in fields or in gutters or scurrying at tube stops i think i do remember when i was a kid my friend had a couple mice as pets + i think i remember at the time thinking that was kind of weird but i may have made that memory up + i also think i remember back in my fiction reading days reading *the rats in the walls* by hp lovecraft but that was rats not mice is there a difference i dont remember + anyway i dont remember how that story ended + as i tried to remember these things i still couldnt stop watching this particular mouse in my room sniffing about while i wondered how it could have fit itself between that tiny wall crack + i wondered then how its bones worked + i wondered then how it came to be that this had all happened on that particular night that same night that all that other shit i told you about happened i mean did it decide to come to me or just end up there through no intention beyond its own biology + also what should i do now try to catch it or leave it to find its own way through my personal private space my escape from out there or do i try to encourage it back the way it had come through that tiny wall space behind which i can only imagine what

Sean Wai Keung is a Glasgow-based poet and performer. His pamphlet *you are mistaken* won the Rialto Open Pamphlet Competition 2016 and was named a Poetry School Book of the Year. He also released *how to cook* through Speculative Books in 2018. His poems have appeared in zines such as *ZARF* and *Monstrous Regiment* as well as anthologies including *The Dizziness of Freedom* (Bad Betty Press, 2019) and *Anti-Hate Anthology* (Spoken Word London, 2019). He has performed with organisations such as the National Theatre of Scotland, Edinburgh Art Fair and Summerhall. Catch up with him on Twitter @SeanWaiKeung or through seanwaikeung.com

Hipster

Stefan Mohamed

First it was *queebly*
computer-augmented inversions
of the traditional folk music
of the Bivek swamp dwellers.
Nobody on the tiny moon
he called home
had heard it.

But then he travelled off-world
and saw its galactic-level popularity

and he went all-in on *cybertechpunkstep*
music made by rogue AIs for rogue AIs.
He collected every
digitally sub-encoded transmission
each with its own unique binary code
in a vast library
half the size of a child's fingernail.

Until the AI revolution
went mainstream

at which point he immersed himself
in *quin bendo*
the four-dimensional psycho-sonic rave projections
of the Solar Quang
a gaseous hive mind engaged full-time
in cosmic hedonism so totally out there
that carbon-based lifeforms
could barely comprehend it
apart from, you know
a select few who just
really GET it, you know.

But then literally *everybody* started injecting
solar radiation to expand their cerebral cortexes

so he spent a few diverting

but ultimately unfulfilling years
bouncing from *shnuppi hardbump*
to *Fomelian anti-jazz*
to *quantum stink*
to *event horizon*
to *gaswave.*

But none of them quite
hit the spot –

until eventually he uploaded his consciousness
directly into the cosmic firmament
so that his spirit could swim in the eternal song
of the deeper universe, absorbing and digesting
its time-dilating chord progressions
and the paradoxical beauty of its ethereal cadences.

Except he soon found himself surrounded
by snooty intangible essences

who had been there
since the beginning of time
and had no truck with *tourists*
jumping on the bandwagon
billions of years in
not when *they* had been around
at the very start
when it was all fresh and new
before *matter* was even a thing
you know
and they turned
the abstract idea of their noses
up and up and up
until he couldn't take it anymore
because really
who needs the revelation
that you can still feel awkward
and ostracised
and inadequate
even when you are merely
the freshly transcended
incorporeal
idea
of a sentient being?

At which point he scrunched himself up with such frustration
that he winked entirely out of existence

leaving behind
a long, high, chiming inter-dimensional tone
that any minute now
some eager kid
at the other end of the galaxy
will be gleefully remixing
as though
they're the first person
to get really mad keen
on the final soul transmission
of a fully sublimated
post-physical entity.

Which, seriously, I was into
long before it was cool.

Stefan Mohamed is a Bristol-based poet, author and spoken word performer. His poetry collection *PANIC!* is published by Burning Eye Books and his novels *Falling Leaves* and the *Bitter Sixteen Trilogy* are published by Salt Publishing. He has also had poems published by Ink Sweat and Tears, Ice Pop Poetry and Cinnamon Press. He can be found at www.stefmo. co.uk or tweeting insightfully about the Vengabus at @stefmoword

Buzzfeed's Top 8 Facts About Space

Mark Gallie

Buzzfeed's Top 8 Facts About Space
The last one will surprise you!

8) You become taller in space.
Turns out you don't have to be called Atlas to hold the weight
of the heavens on your shoulders.
Up to an extra 5cm taller they say in the inky black
but one day I will have to come to terms
with the fact that even if I cast aside the FORCE OF THE UNIVERSE
I will never reach 6ft tall.

7) The footprints on the moon will be there for 100 million years.
Without the wily involvement of those pesky elements
we deal with down here called our atmosphere (for now)
those footprints will never vanish and
that's the type of lasting imprint I want to leave behind
but not as physical.
Given the choice my footprints would be between
"the guy that was always there when you needed him" and
"the guy that tried to become a duck".

6) One day on Venus is longer than one year.
I think I can relate –
because like the days following the dream job application,
the minutes stuck in a lift with your ex, or
the moments after a STI diagnosis, some periods of time linger
WAY LONGER than they should.

5) There are an estimated 500,000 pieces of 'space junk' drifting in our atmosphere.
From rocket parts and refuse
to sloppily dropped spanners, this
makes me uneasy about our future as
awesome space pioneers, adventurers, or
even just as inhabitants when we are already responsible for a littering problem.

4) In 1977, a craft was launched that has now reached interstellar space.
Aboard it carries a golden record to act as the time capsule of Earth.
From human greetings to 70s music,
I feel like first contact is not going to be as the film industry would have us believe.
Less invaders wielding laser beams and

more confused tourists wondering why there is a lack of Peruvian panpipes
and a surplus
of cat pictures.

3) There is an invisible boundary 100km from here that marks the beginning of space.

That means if I took a car and
drove straight up
I would hit space faster than I could get from Edinburgh to Glasgow
providing the traffic between here and THE VOID is good.

2) Nobody knows how many stars are in space.

The sheer size makes it impossible to predict just how many there are, and I don't know how to deal with that.
I can't fathom something an hour's drive away
being so super colossal that the term SUPER COLOSSAL
doesn't do it justice!
So for now I'll stick with this space,
these people, and I'll take it slow,
just a small step at a time.

1) ARRRRRGGGGGGHHHHHHHHHH!!!

Told you.

Mark Gallie is an Edinburgh-based spoken word performer, actor and writer. Mark is part of production company I Am Loud and has performed nationally and internationally as a featured poet. Mark also supported Shane Koyczan during the Scottish Leg of his 2019 UK tour. He is a producer to monthly showcases, podcasts and writes scripts for commissioned work.

Stephen Hawking Throws a Soiree for Time Travelers But This Time, They Show Up

Rikki Santer

First the professor coordinates space and time—
52 degrees, 12 minutes, 21 seconds north
0 degrees, 7 minutes, 4.7 seconds east—
another sumptuous recipe from geography,
which easily rivals sliced bread.
On the counter, his Foodini printer gleams ready
for any 3D Frankenpastry that may be desired.
As time breaks its moorings, Simpson cartoon
clouds part and light seeps into full moon.
Worm-holed planks of a buffet table snap into place,
bouquets of jewel-toned balloons punctuate the drawing room
then in response to the muscular lilt of his grin
doors of an interplanetary elevator open
and his guests, having crossed borders
through a gizzard squeeze, leave snail
trails on the silk brocade while packing
deus ex machina in their pockets.

Hawking's pendulum clock, steady and sure,
ticks out a town for these tourists—
togas and tuxedos, low-backed dresses
and black turtlenecks, silver spandex,
and light-weight spacesuits bio-engineered.
They sample canapés of flummery
topped with quantum foam,
five loaves and two fishes,
a platter of forbidden fruit,
a 12-tiered cake studded
with Marie Antoinette's pearls,

Cleopatra's stuffed pigeons,
Vitellius's flamingo tongues,
armadillo fricassee especially
for Darwin, breakfast eggs
with porcini mushrooms
if Einstein decides
to sleep over.

As each moment buds into another,
a troupe of grandfathers and grandmothers
wield paradox shields
as they linger and swell
around a holographic sundial.
And over there, H.G. raising
a Waterford flute of champagne,
steadies a nod to Stephen
with his mustache.

Rikki Santer has appeared in numerous publications including Ms. Magazine, Poetry East, Margie, The Journal of American Poetry, Hotel Amerika, Crab Orchard Review, Grimm, Slipstream and The Main Street Rag. Her work has received four Pushcart and three Ohioana book award nominations, as well as a fellowship from the National Endowment for the Humanities. Her seventh collection, *In Pearl Broth*, will be published this spring by Stubborn Mule Press. www.rikkisanter.com

In Baydhabo

(After Mahmoud Darwish's In Jerusalem, with a line from)

Asmaa Jama

Here sitting across from me is a second
reality, the Chinese, amongst the small things,
sent a photon into space. I mean to say in
Baydhabo I noticed a cousin of mine trip
between ribs, and land in space. Or the space
between his ribs filled with dates. I mean to
say, I was afraid of going outdoors, where
the acid might spray on my face. How can
I tell which land is my land, this land where
everything lands amongst the pre-wrapped
cadavers, or that land where in amongst the
spitting fats, a photon was transported and
was found in space. Sitting before me is
a version of some unnameable ancestor. I
mean to say, I cannot pronounce his name and
this face is a sorrowed face, and my skin
becomes its own border. And my
father's father's father was wrapped in a cloth
they no longer make. And the factories of my
father's father, have started to break. And nothing
grows from the broken place. And my father's legs
snapped open one day, and nothing grows from the
broken place. And a photon was transported and
found in space, and I have always dreamed of an

unclaimed land, that I could forever never claim, land

where my passports cling to me, I flee from them

and when I cling to my passports, they flee from me

and freedom must be a state of mind and not a place.

And a photon, a whole actual thing of light, that was

caught on Earth, was teleported, yes like in the movies,

into space. And if that's possible, then why can't I live

and pray on some distant moon, that will go unclaimed,

the mosque being wherever I make it. Or why don't I

make a mosque and a minaret out of the Moon, and sit

inside it, the carpets crescenting my feet, and never

pray, or watch the others pray, and feel as a guiltless as

a child. And not fear the doors unbarricaded, and not fear

that in this mosque on this carpet, my life ends, unbarricaded

from everything alive, and I depart unwuddhu'd and complacent,

and I depart with nothing to prove, other than I was alive.

What do you want from me, you killed me,

and I forgot, like you, to die.

Asmaa Jama is a poet and visual artist based in Bristol. Asmaa uses her artistry to speak intimately and honestly about her identity as a Danish-born Somali, and her experiences of being a third-culture kid in the UK. Her work has been published in Popshot Magazine and Paper Swans' *Young Poets' Anthology* and she was commended by the Young Poets Network. As a visual artist, she has been featured online by Dark Yellow Dot and, as a writer, she has contributed to Rife magazine. In 2019 she participated in Level Up, a course for talented poets run by Blahblahblah and Milk Poetry. Instagram: asmaa_floats / Twitter: asmaa_floats

Take Care, I Love You

Helen McClory

Sections in bold taken from the contents for the Wikipedia article on the Fermi Paradox, which examines the contradiction between mathematical probability that suggests alien life should exist in our universe and the lack of any evidence that it does.

5   Hypothetical explanations for the paradox

A poem about loneliness

5.1   Extraterrestrial life is rare or non-existent

You wonder about other people

5.2   No other intelligent species have arisen

They seem to manage and group together

5.3   Intelligent alien species lack advanced technology

You stare at your phone to keep the room from collapsing

5.4   It is the nature of intelligent life to destroy itself

You're not sure how smart or stupid you are, it's going along okay, mostly

5.5   It is the nature of intelligent life to destroy others

At work you offer your colleague a mince pie but they are distracted

5.6   Periodic extinction by natural events

You take your time walking around the supermarket, it is bright and busy

5.7   Inflation hypothesis and the youngness argument

You play *Witcher 3* when you get home for six hours and it's much more beautiful than real life

5.8   Intelligent civilizations are too far apart in space or time

You can go on holiday by yourself, that's fine, you're saving up for Italy in the new year

5.9   It is too expensive to spread physically throughout the galaxy

Your mum said she was glad you were doing so well

5.10   Human beings have not existed long enough

It's so hard to know what to do

5.11   We are not listening properly

You're not able to tell anyone

5.12   Civilizations broadcast detectable radio signals only for a brief period of time

You like the idea of Snapchat, but there's nobody who'd do it with you

5.13   They tend to isolate themselves

The last time you broke down and actually talked to your mum about how you feel all the time she said 'where did I go wrong?'

5.14   They are too alien

You might look for a new job, you might fit in better somewhere else

5.15   Everyone is listening, no one is transmitting

You check your texts even so

5.16   Earth is deliberately not contacted

Of course you're going to say, 'it's all my fault' because you always make it about you

5.17   Earth is purposely isolated (planetarium hypothesis)

Looking out the window at the old man tottering along on the icy pavement with his wee dog, wondering if he has a family

5.18   It is dangerous to communicate

You very nearly told the cashier 'take care, I love you', and you'd have meant it

5.19   The Simulation Theory

It has to be better than this

5.20   They are here undetected

You've joined Twitter but you don't really know how it works, you're trying

5.21   They are here unacknowledged

In your new profile you've just put what seemed right: take care, I love you.

'Take Care, I Love You' is taken from Helen's short story collection *Mayhem & Death* published by 404 Ink.

Helen McClory lives in Edinburgh and grew up between there and the Isle of Skye. Her debut, *On the Edges of Vision*, won the Saltire First Book of the Year 2015. She has since published *Mayhem & Death* and *The Goldblum Variations*, following Jeff around the known (and unknown) universe, which will be published by Penguin Books in the US in late 2019. There is a moor and a cold sea in her heart.

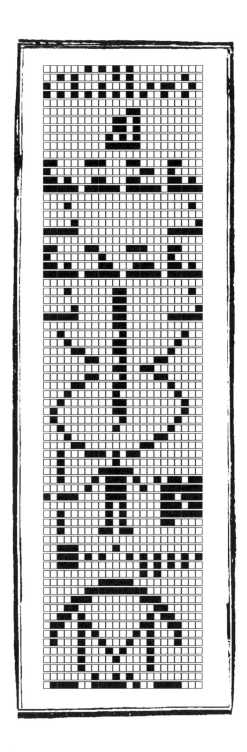

FICTION

Jeff Bezos Does Eurovision

Annie Summerlee

No one expected Jeff Bezos to make it to the Grand Final of Eurovision, but here he is, backstage, bottle of water in one hand, microphone in the other. Mars is belting out high notes, the crowd is roaring, and Jeff, representing Pluto, is up next.

His ears are ringing. Sweat coats his bald head, running behind his ears. Even his hands glisten. Someone —a backup dancer— is patting his back, squeezing his bicep, saying: you'll crush him. Bookies still think Mars is going to win. They've taken the trophy home twenty-six times, and it's no surprise that the contest is here again after last year's victory.

Pluto is yet to win. They've attempted, year after year, to find a loophole leading to disqualification, but Eurovision is so deeply ingrained into Inter-Planetary law that they settled, fifty years ago, with sending war criminals to compete. Being humiliated at the quarter finals after years of space travel is considered sufficient penitence for the scattered few trapped behind bars.

Bezos was selected to represent Pluto nine years ago and has been on Mars for nine months preparing himself. "If you win," he hears his backup dancer say. "You'll be a free man, Jeff."

He wonders if that's true.

Jeff used to be President and Supreme Leader of Pluto. He had the biggest, warmest house, and his people loved him dearly, until that one fateful commemoration day, when a Martian journalist discovered he wasn't the real Jeff Bezos, colonizer of the outer orbits of the Solar System. The real Jeff Bezos had suffered from stage fright before his ship was due to leave Earth, and sent a look-alike instead, ordering him to be charismatic and ruthless and take over the Universe, a task at which he failed, with the exception of Pluto.

And so the people of Pluto locked not-Jeff Bezos up and decided to send him to Eurovision. His people chose his song for him, which just so happened to be an accidental masterpiece of space-post-pop. At his first press conference, a newspaper from Venus asked if he believed the hype around his song, and only then did Jeff realise there was hype around his entry. At the quarter-final, he had the crowd singing his lyrics back at him. At the semi-final, with big guns like Mars and Saturn's moon Rhea, he expected to be eliminated. But he wasn't and now bookies have him in second place, just behind Mars.

Mars lets out one last note, microphone held high above his head. The crowd goes wild and Jeff's backup dancer says, "You've made it this far, Jeff. Jury final was a breeze. If you win, they'll promote us back to planet again."

The crowd — almost half of them look human — is cheering his name, and he struts out with his five backup dancers to the centre of the glass stage. "I believe in you, Supreme Leader," the backup dancer says.

"For Pluto," Jeff says, and *For Pluto* all his dancers, state criminals the lot, reply.

JEFF JEFF JEFF, chants the crowd, and he tries to dry his sweat, taking a deep breath.

"You've got me pissing in a bottle." Just as he sings the first line, his sweating palms cause the glittering mic to slip through his fingers. His dancers keep dancing, but Jeff can't see through the coat of smoke on the floor; he's down on his knees, cameraman following his panicked eyes. Backstage, Mars cheers. He finds the microphone after fifteen seconds, but as he tries to keep singing, all he remembers is *pissing in a bottle*, so that's what he sings, until his dancers glide offstage and he rushes behind them.

They're going to let him sing again. He wishes they hadn't said that, that instead of technical issues they'd said human incompetence. He dusts his hands with talcum, and his backup dancer — a man that Jeff (as Supreme Leader) sentenced to life in prison for the murder of eighty-two pensioners — looks him in the eye and says, "Don't disappoint me, Jeff."

They call for Pluto again, and the crowd, so compassionate, chants his name once more. He takes off his rocket-shaped costume, and quickly puts it on his dancer, hands him his microphone and says to him, "Pluto better win Eurovision."

(Mid-song they're asked to stop: changing a performer last minute during the Grand Final isn't allowed, and Pluto is suspended from participating for fifty years; the next nine acts, already en route, turn back, their chances at stardom soiled).

Annie Summerlee lives in Catalonia with her photogenic cat. She has published short stories in literary magazines and anthologies, writing in Catalan, English and Spanish. She is currently working on her first novel.

An Inventory of Hidden Spaces

Georgia Dodsworth

A cavern under the earth.

Distant groaning, an elemental screech, growing louder. Two white eyes pierce the darkness, widening.

The train is moving a mile a minute; the passengers inside sway on the balls of their feet, shoulders shaking. Commuters, tourists, children, jolted across the underside of the city, staring into nothing.

The space inside the reusable coffee cup in the grip of the woman in the green coat holds around twelve liquid ounces. If you were to take the cup from her hands and remove its lid, steam would not rise to your face, but the smell of... is that–?.

The odour of her morning vodka is of extreme concern to the woman in the green coat. Years later she is still of the conviction that if she leaves it undiluted, or lets her cup get too close to another passenger, they will know, they will all know and stare. So she cuts her drink with lukewarm cola every day and brings it to her lips very deliberately, though her fingers always shake. On crowded mornings she places her thumb over the hole in the lid, lest a tell-tale fume escape. The fear of being knocked to the ground and having her secret spill across the sticky floor is almost more than she can bear.

And yet it is too painful to be sober, makes her antsy, makes her sick. But more than that: when she holds her secret space inside her hands every morning, it is a tiny thrill to know she is the only one who knows, to know she is brave and strong and bold enough to take this risk. She allows herself another sip and winces: give me strength.

She wonders if the sad-faced girl standing by the railing is looking at her. She is not; her gaze passes through the woman in the green coat and beyond. In her mind the sad-faced girl is sitting with a boy she has not spoken to for weeks, drinking and laughing on the roof of his house. Or rather, he has not spoken to her. To think of this boy brings her pain, but every morning on the train she thinks of him nonetheless.

Why did she say what she said at that party? The space in the carriage shrinks to that small, dingy kitchen every morning, his reddened face glaring at her from across the room, that look of betrayal that makes her feel ill.

Tears form, and she is back on the train. If she cries, the disparate focus of each passenger will shift towards her, in silent frozen pity. She sucks in a breath and pulls her blazer tighter, clutching her sensible work bag in front of her stomach.

The sad-faced girl has not yet noticed that the space behind the bag has begun to expand, millimetre by millimetre, cell by cell. She has no suspicions yet, and won't for several weeks. She will in fact realise on this same train, on the same kind of dull morning, just as she has started to forget about the boy on the roof.

She glances up. The businessman in front of her leans forward, arms rested on his legs, a slim, threadbare book in his hands. She strains her eyes to focus on the top of the page – *The Sentinel*. She leans back; she doesn't know what that word means.

The businessman has only a page left, but he has been reading the same sentence for the last eight minutes. His mind is on the long, drab day he is hurtling towards, a day he wishes he could sail right past. No use; he slams the book shut, sits back in his seat. He is unaware that earlier this morning, in the infinitesimal space between those last two pages his four year-old daughter placed a drawing of a tree, signed and addressed to him in her shaky scrawl. He will not realise until the train journey home, when it will slip from the pages and, laughing, he will see her toothless grin and big blue eyes before him, his own little girl.

But for now, he is in the foulest of moods. He glances around the carriage and imagines grabbing the head of each passenger with both hands and shaking it, one by one. This violent impulse unnerves him, but the thought melts away to the next: he suddenly remembers a book he read as a child, one which detailed dubious stories of unusual events. Tales of frogs dropping from the sky, a boy whose lung became wrapped in a vine after inhaling a seed, a man born without a brain.

He thinks about this last one from time to time and isn't sure why. He remembers the author's sincere reassurance that the story had been proven, that if you held a light to the back of the man's skull, an orange glow would suffuse behind his brow.

The businessman glances once more around the train. If you could open a person's head, what would you see? He glances down. If you picked their pockets, what would you find?

A pack of cigarettes tucked at the bottom of a backpack, empty save for an engagement ring. A racy paperback concealed inside the dustjacket of a more innocuous tome. A ten-pound note hidden in the heel of a boot – for emergencies, of course.

Next platform: Central.

The train slows, and the collective rises. After an age the doors glide open; two spaces merge, exchanging their passengers one by one, before separating once more. Days begin, days end.

The train slides off again; a thousand hidden worlds move onwards.

Georgia Dodsworth is a 24 year-old Applied Linguistics student living and drinking in Edinburgh. Originally from Middlesbrough, which isn't as rough as you've heard. she divides her time between the cinema and the library and is trying to write as much as she can.

The Nimble Men

Mark Fleming

"What have you been told about personal calls, Calum?"

I watch Kennedy's lips moving but hardly take it in. I'd been staring at Emily's text for so long the letters are imprinted ghost-like on my eyes.

Cal. I was trying to explain this morning before you rushed out. I still want to go on holiday, but I think we should go as friends. We can chat later. Emily X

How incongruous is that kiss? Like an inmate being informed the date of his lethal injection, then a choice of his favourite last meal. I shove the phone back inside the sporran, fingers wrapped around it. Squeezing until my palm aches. I try holding Kennedy in just as intense a glare. But my stare relinquishes, as it always does.

"It was important," I mumble. "But I've switched it off."

"Good. Serving customers is what's important. Keep Scotland's history flowing, Calum."

He strides towards the advance guard of the next geriatric coach party. I note the accents of the closest bunch. New Zealanders. Several wearing All-Blacks rugby tops. Their version of the haka would look more like the *Thriller* video.

This was supposed to be my last day manning this kiosk at the Culloden visitor centre. I was due to start packing for Thailand tomorrow, before heading over to Aberdeen airport on Saturday for the first leg of a year's backpacking. But Emily's bombshell has derailed everything. All the arrangements. New clothes. Currency. Vaccinations. Mosquito repellent. Months of planning cancelled in a few lines of text.

Insisting we can still travel half-way across the world together because we'll remain friends? Travel companions? Maybe this is one of her episodes. It wouldn't be the first time there's been a wobble in our relationship. That time I went down on bended knee in a restaurant and she simply guffawed with laughter. "In time," she said. I think of sleeping in the same room and doing just that. How would a platonic thing survive each time we got melted?

My phone vibrates. Kennedy is elsewhere so I check it out. Another text.

I have to explain something, Cal. There's a guy. Davie. From Melbourne. Games designer. We met online. He's coming to meet me in Bangkok. You'll like Davie. I think you'll get on just great. He's into history. You can tell him all about your Highland battles. He could get inspiration. He designs war games. We can be mature about this, Cal. When Davie joins us, you can book your own accommodation. By then, you'll probably have met someone anyway. The beach parties are awesome over there, Cal. Lassies from every corner of the globe, in bikinis, eckied up. Your fucking guaranteed, mate!! We can go out in foursomes. Get chonged watching the sun setting over the Andaman Sea. X

Joining them? My sincerest wish would be for Emily and her virtual lover to be arrested by customs after being forced to smuggle cocaine into Indonesia.

Closing my eyes, I bunch my fists. Try breathing evenly, although I feel every muscle trembling. Opening again, I stare at the souvenir displays. Scottish terrier fridge magnets. Lion Rampant mouse mats. Loch Ness Monster cuddly toys. Braveheart keyrings. Scottish culture reduced to meaningless trinkets, the equivalent of Sitting Bull performing horse stunts with Buffalo Bill's travelling circus.

Two middle-aged men waddle from the cinema. Their accents remind me of *The Sopranos*. New Jersey. They hover, fingering postcards, then paperweights bearing clan surnames. I clock regimental tattoos nestling amongst the wizened white hairs on one guy's tanned forearms. Not old enough to have survived Normandy nor the Pacific, but too old for Vietnam. Korea? They're joined by respective wives in leisure suits.

"Just think on it, Melv?" says one of the housewives of New Jersey. "Two thousand men died out there, out in that field, just to decide who was the rightful British King."

"Unbelievable, Jen. Bonnie Princes? Young Pretenders? What is this, Hans Christian Andersen?"

My mind flits back to Emily. How can you break up a real relationship based on a connection with someone you've never met? What if Davie isn't Davie? What if Davie's the fake persona of a 60 year-old weirdo from Milton Keynes? Even so, when I think of the situation, he's real enough to me. Maybe they've Skyped when I've been out. You think you know somebody, but if you find out you actually *don't* fucking know them, you can imagine anything.

I gaze around. My parents used to take me to the old visitor centre as a kid. The redcoat commander, the so-called 'Butcher Cumberland', who ordered his men to give no quarter to the wounded on the battlefield, was portrayed as a rotund mannequin with rosy cheeks. There was also a dressing-up box with redcoats for his soldiers. Jimmy wigs for the Jacobites.

My phone vibrates. I snatch it. Her again. A call this time.

"What the fuck, Emily?"

The Americans watch me. Kennedy is hovering again, furious.

"Cal. Don't be so immature. We always agreed we could see other folk, didn't we? Christ, we're nineteen. We're far too young to get serious. There's a whole world out there. Anyway. You'll get on with Davie. I *promise*.

Cal? Speak to me, Cal. I'm sure you'd rather I was honest? You're not gonna be a baby about this, are you?"

I toss the phone across the room. It cracks against a pre-1746 map of the Highlands. All the clan territories still there in a multi-coloured patchwork.

I snarl at the nearest woman. "That cartoon Nessie on your t-shirt? Tell me which is more fucking farcical? That plesiosaurs could have survived the extinction of 99% of the dinosaurs 65 million years ago, continued to survive, right through the Ice Age that carved Loch Ness from the mountains, then live for another 13 thousand years? Or the notion that dinosaurs wore tartan fucking bunnets?"

"What did he say to you, honey?" her husband spits out. "Sometimes I find the local accent is just goddamn impossible."

"Search me," she replies. "It's supposed to be English but it's anything but."

Kennedy is rushing over, combover flapping, jowls wobbling. He sometimes reminds me of that Duke of Cumberland dummy; at this moment, all that's missing is the white wig.

"What on earth has got into you, Calum?! You're not just demob crazy, you're crazy full stop!"

Said with a smirk. He's been gunning for reasons to get rid of me before my last day and I've presented it, with loan shark interest.

"This boy's gone berserk," pipes an American voice.

Kennedy stands between the tourists and me, raising his hands in apology. I elbow my way past.

'Calum?'

I stumble against a display cabinet. A dragoon's pistol and a bayonet unearthed from the soil. I elbow a crack in the glass panel. A woman squeals.

"Calum!"

Next thing, I'm outside, stumbling across the moor. I can't think where I want to go, only what I need to leave behind. I run by the red flags indicating the government army's deployment on the morning of 16 April. Soon my legs are aching. Lungs struggling for air. But I keep on. Arriving at the blue flags, the Jacobite position, I halt.

A skylark trills. Clouds like vast cobwebs smear across the powder blue. I gaze back at the red flags, imagine a young rebel, his terror-stricken vision darting from the cloud formations to the scarlet uniforms formed in disciplined ranks. He would have gripped his claymore with white knuckles. Preparing for the order to rush towards those musketeers and cannons primed with grapeshot. Offering his life for Prince Charles Edward Stuart, not through any deep-rooted political leanings, not necessarily through religion since the Jacobite army contained as many Protestants as Catholics, but simply because his Laird, his master, had aligned himself to this cause.

As for the government troops, fighting for a government only 3% of the population were entitled to vote for, and for King George II, a monarch whose native language was German? They'd be watching the rebel army with a mixture of trepidation, anticipating the feared Highland charge, but also keenly aware of what volleys would do to men struggling over heavy ground.

Perhaps this rebel was praying for a horseman to appear on the horizon, through the dawn mists, brandishing a white flag, bringing news that the Houses of Stuart and Hanover were to be united by some marriage. That was how alliances were formed back then. Weddings. Or killings.

Reaching into my uniform's plastic sporran I snatch my lighter, spark the doob I'd prepared for my walk to the bus stop after my shift. I exhale towards the clouds. Now I delve into the pouch for the hip flask. Uncapping this, I slug at the whisky. Peering around the moor, I drain the contents in a succession of greedy mouthfuls. Leaning back in the undergrowth, I watch the clouds drifting in a fantastical procession. There's the white Ferrari I'll own one day. There are the Outer Hebrides, joined to a reversed Italy. I follow a jet trail, 12 kilometres high, until its wake fades. The ground rocks gently. My eyelids turn to lead.

* * *

Twilight is enveloping the moor. Beyond the last clouds, the first glimpse of Venus and a dusting of stars. As the surrounding terrain dissolves into shadows, I switch my attention to the infinity of space. Ten minutes later I can pick out the Plough, the Seven Sisters; brightest of all, the Dog Star. Lighting the joint again, I shift the red dot between the points of light, faster, creating dazzling patterns. With each passing minute, further reaches of the infinite majesty of space are revealed. Hundreds of stars. Thousands.

Gradually I notice curtains of green and blue light shimmering near the horizon. The Northern Lights. *Na Far Chlis*, as the Highlanders would've referred to them. The Nimble Men. I visualise the fighter again, the Friday night before the battle, 300 years ago. Tartan shawl wrapped tightly around his shivering frame, fingers hovering by a campfire, sucking on a clay pipe, gazing up at these stars. Inventing his own constellations; seeing wolves, leaping salmon, pouncing kestrels. Creating his own mythology as soldiers on any battle's eve have done for millennia. Wondering if the Nimble Men might appear, gambolling across space. Placing everything in perspective, the glory of the heavens versus the ridiculous sectarian squabbling of men on this tiny planet. Most of all, wondering if he'd be alive to lose himself in space the same time tomorrow, after it was all over and the air was thick with the scent of blood, the screams of the maimed; seeking the Nimble Men leaping over these skies in their eternal dance.

Flashing lights by the centre. Above, an aeroplane blinks over the Moray Firth. Taking a final, lingering drag I notice torchlight by the red flags. The sound of barking dogs makes my hair bristle. I think of Emily snorkelling through the South China Sea, a tanned figure floating close by, ripped muscles dragging through the blue waters.

The wind stirs the blue flags above me. Black flags now. They flutter and snap. Torches carve swathes of heather from the gloom. One beam sweeps by. Catches me. Others home in. Flicking the roach away, I take a deep breath. Begin marching up the slope towards them. Quickening my pace from a stroll to a jog, my heart drums in my chest. Breaking into a sprint, I lift my face to the stars. A deep voice bellows. Torchlight envelops me, as if I'm being ambushed by paparazzi.

Shrieking like a proud clansman. *"Kennedy? You called the fucking busies?"*

I don't even remember snatching the bayonet from the glass shards. As I brandish it in the night air, I feel matted blood between my fingers and the hilt.

The crackle of a taser. 50,000 volts. I slump backwards, limbs jerking. Above the battlefield, space is a manic whirlwind, as if galaxies are being born. The Nimble Men won't stop.

Mark Fleming is an Edinburgh-based writer. His short stories have been published in outlets such as The Big Issue in Scotland, football fanzines, *Shorts*: the Macallan/Scotland on Sunday short story collection, and *The Picador Book of Contemporary Scottish Fiction*.

FEATURES

NEW FRONTIERS WITH NASA'S ADRIANA OCAMPO

by Katie Sharp

Have you ever glanced at a clear night sky...

...at the two-thousand-odd stars observable from Earth and pondered the countless questions that the twinkling pinpricks of light can inspire? If so, you are not alone in marvelling at the vastness of the universe. Humans have been interested in the night sky since before they could write their questions down. As a species we have grown and evolved but we are only now becoming acquainted with our own skies. It takes an extraordinary person to turn their curiosity into a career, and that is exactly what Adriana Ocampo, Science Program Manager at NASA, has done – in spectacular fashion.

Adriana has taken her childhood dreams and turned them into her lifelong passion, helping to shape humanity's journey to the stars. When it came to exploring the topic of space, there was no one more fitting to speak to, and so join us as we journey to NASA Headquarters and beyond.

THE BEGINNING

"From a very early age, I was just fascinated looking at the skies and those little points of light and wondering what they were. Were there people like us there? I used to go to the rooftop of my house in Buenos Aires, Argentina, where I grew up. I was born in Columbia but grew up in Buenos Aires and used to dream about travelling in space. I used to steal the pots and pans from my mother's kitchen and things from my father's electronic workshop and build make-believe spacecrafts on the rooftop of the apartment.

'Queen Adriana', as we've dubbed her
Credit: NASA/Aubrey Gemignani

"Space intrigued me from a very early age and I was fortunate enough to have parents who encouraged me to dream big and use my imagination. I was given a family motto: *Work hard and your dreams will come true.* I was taught that study and education was the path for enabling your dreams and space was my dream. I'm a child of Apollo because on July 20th 1969, the launch became a key moment for me. We were all crowded around the only TV that was in the neighbourhood, watching that amazing day that changed human history, so I knew that dreams could come true.

I WROTE A LETTER TO NASA. I DIDN'T KNOW THE ADDRESS, I DIDN'T KNOW ENGLISH, I JUST WROTE TO THEM

"My belief in my dream was strengthened especially after I wrote a letter to NASA. I didn't know the address, I didn't know English, I just wrote to them in Spanish telling them I wanted to be part of their team. I didn't have their address, I only put 'NASA, United States' on the envelope and remarkably enough, thanks to the incredible postal service across the world, it actually arrived. Somebody at NASA took the time to answer me which was the greatest treasure because it confirmed for me that dreams could come true."

THE ROAD TO NASA

"The first thing that I wanted to know [when I landed in the USA] was where NASA was. Eventually my parents, by chance, were able to move to South Pasadena, which is close to Pasadenia, where the Jet Propulsion Laboratory (JPL) resides. The JPL sponsors a JPL Space Exploration Post which is where 'explorers' get together and share their common passion, space and space exploration. They came recruiting at the high school I was at, so I immediately volunteered and the rest is history.

"That was such a tremendous experience, and to this day the JPL still sponsors students and allows scientists to volunteer to mentor these students, where they get special lectures and hands on experience. In our case this was to build a transmitter that could communicate with another satellite: the transmitter which was eventually able to communicate with a weather satellite. It was an extraordinary experience because I was a shy young girl and they asked me if I wanted to lead this team. I said, 'Me? No. What?' but a mentor who I still keep in touch with, a past engineer at JPL, Mr Michael Kaiserman, said, 'You can do it, take the position,' and so I became the leader of the team. It was a very vigorous experience because we truly learned how NASA works. The explorers that were involved in that experience have become very much involved in space careers – it was a truly life-changing experience.

"After that, I asked the JPL to hire me part-time, to do anything. I wanted to continue being engaged because I was going to university, so before I finished high school I started working in the very lowest JPL position at NASA which is a technical aid. NASA became my second home. I learnt from working the trenches all day all the way up to the Lead Program Executive for the New Frontiers Program."

NEW FRONTIERS

"This is the only agency program that is led by principal investigators and is where NASA gives an opportunity to the scientific community to propose ideas of mission conceptions to explore the solar system for specific targets, as dictated by the National Academy. Each of these missions has over a billion dollar budget so it is a great responsibility and has been an incomparable experience.

"Part of my portfolio includes the New Horizons mission which went to Pluto; most recently we have been exploring a new object in the Kuiper Belt which is called 2014 MU69, but unofficially we call it Ultima Thule which I think translates to 'The Last Outpost'. The second mission in my portfolio is Juno in the New Frontiers Program. This is a mission which is currently, successfully, orbiting the largest planet in our solar system - Jupiter. It is, for the first time, using solar panels as the source of energy and has been providing incredible insights into how planets are formed and the role that Jupiter played in the formation of the solar system. It has even been giving insights into life on our planet because it has a connection to the water molecule and hints at how water may have arrived on Earth. This is an unprecedented set of findings revealed by the Juno Mission.

"We also have the OSIRIS-REx mission which is currently orbiting the Bennu Asteroid. Bennu is a very interesting asteroid because it is rich in amino acids and organic material which makes it very exciting to explore. For the next year and half we will be mapping Bennu in harsh detail and eventually we will find where to extract a sample which will be returned to us in 2024. Finally, the fourth mission which is currently under completion – this will be a mid-sized mission, very ambitious, with very targeted scientific objectives for exploration."

A photo of Bennu taken by OSIRIS-REx on June 13, 2019
Credit: NASA/Goddard/University of Arizona/Lockheed Martin

A DAY IN THE LIFE

"Day to day, communication plays a key role in everything we do. It doesn't matter how scientifically or technologically complex the work may be, if we don't communicate well and work as a team it is hard to carry a project to a successful end. I try to maintain the communication level going on within the team, between the Principal Investigators, science team, and the engineering team. I am also currently working in a new mission which was selected two years ago as part of the Discovery Program called Lucy, which will be going to explore the Trojan family of asteroids that orbit the front and back of the orbit of Jupiter. Lucy is a very complex mission aiming to assign a trajectory to these asteroids, which is an ambitious task. The mission will basically visit seven objects, seven asteroids, with one spacecraft.

"When a mission gets selected you have some time – the Principal Investigator has never done this before and has such a big responsibility. They have this big team of scientists and engineers who are all extremely talented, and you help them. Part of being a Programme Executive is leading and managing by influence to make sure that the mission fulfils all the agency requirements, the scientific requirements, technical requirements and help shepherd the different phases that a mission has to go through. It is also your role to manage the technical reporting to eventually carry it to launch and then on to a successful mission accomplishment.

"It is a combination of knowing what the science is, knowing what the technical challenges are, handling risk every day and maintaining fluid communication. Open door communication and communication that is not based in blame is important. If something goes wrong, we can learn how we can do it better. We think, 'Okay, where do we go from here?'

"I also get to do a lot of fun things. With New Horizons, I got to help organise an astronomical campaign to do the locations of the 2014 NU69 Object. I worked with Argentina, South Africa, Columbia and Senegal to help determine the size and shape of this object that had just been discovered in 2014. The spacecraft was going to fly over it in January 2019 so we needed to mitigate the risk to the spacecraft and maximise the science return. By doing these five astronomical campaigns, with locations of 25 telescopes and with a science team of about 60 people, it was unthinkably challenging. It had never been done before. We had to orchestrate this to the miscrosecond to ensure all the telescopes would be pointing at the right place, in the right location in order to capture this event. We had to time it impeccably, just as the object NU69 was going to cross in front of the star, right to the millisecond. And we did it. As a result, the spacecraft was able to fly close to the object which brought us breathtaking data.

"I also shepherd agency activities to do with Venus and Venus exploration. A lot of that involves working with our international partners which include Russia, the European Space Agency, and the Japanese Space Agency, so I have a very big and varied portfolio that has been a lot of fun to manage. It's not only desk work, I also get to go and do fun things in the field.

> WE BELONG TO A MUCH MORE COMPLEX SOLAR SYSTEM, A MUCH MORE COMPLEX, EVER-EVOLVING UNIVERSE. ... EARTH IS A CRADLE, WE NEED TO LEARN TO GO BEYOND THE CRADLE.

SO... WHY EXPLORE SPACE?

"An argument I frequently hear is, 'Why do we spend so much money on space exploration where we have so many challenges to solve here on Earth?' I'd answer that, not only does space exploration result in numerous jobs, space itself provides an element of inspiration that is invaluable and so often overlooked.

"It gives inspiration to the nation, to us as humans, that we are more than just a species that lives on this planet. We belong to a much more complex solar system, a much more complex, ever-evolving universe. If we are truly going to survive as a species, it has been said many times – Earth is a cradle, we need to learn to go beyond the cradle to really evolve and become interplanetary and eventually an interstellar species. So, in essence, it is in order to ensure the survival of the species and how we can evolve into something better."

TOP TIPS

"The number one thing is to follow your dreams and never give up. This is at the heart of everything. If you are true to yourself and true to what your passion is, you will become the best at it and the job will not be a job, but it will be a fulfilling employment.

"NASA employs everyone from engineers and scientists to doctors and lawyers. There are architects and design specialists, and there are people with computer expertise. There really is no discipline that is not involved in space.

"The other key thing is that we are at a breakthrough era. This is the era in which we are creating the road map we will use to leave our planet. To not only return to the Moon but to stay there. This data will act as a stepping stone to having a human presence on Mars that could eventually lead to a permanent presence on Mars. Space has already touched all our lives, and this is something the next generation will have to be prepared for. The things that we take for granted, the things we use every day - I can guarantee if you trace the origins of around 80% of them, you'll find them rooted in space exploration. The next frontier does not exist in the future - it is here right now.

"Take it back to your dreams – whatever that passion is, follow it. Don't let anybody divert you. Make sure that if you get 'Nos' and failures along the way that you take them as the way to get to the 'Yes' that you need. This is one of the biggest lessons I have learned working at NASA – we grow from anomalies, challenges and problems. I learn how to do things better, I learn how to get different, more creative and innovative ways to reach that objective. Never give up and always follow your dreams."

A BONUS FINAL QUESTION... WHAT'S THE MOST EXCITING THING ABOUT WORKING AT NASA AND DISCOVERING OBJECTS IN SPACE AND EXPLORING PLANETS AND MAKING DISCOVERIES THAT CAN IMPACT OUR UNDERSTANDING OF THE UNIVERSE AND THE WHOLE FUTURE OF HUMANITY...?!

"Everything! It is hard to believe every day when I come to work the stuff that I get to do. I love that we are always thinking. I love our language. Every day we talk about Mars, we talk about Jupiter, about exploration, about how we are going to go to the Moon. We talk about how humans will be working with robots. I enjoy our innovation, and that there is a whole branch of the agency that uses the patents produced by NASA.

"Carrying a mission concept from an idea all the way to the launch pad and on to its target, there is nothing like that. I still get chills seeing New Horizons in the launch pad after so many years of being in concept, of the mission being cancelled and restarted again – the same thing with Juno. Every mission is like a new child is being born, a new set of options and knowledge that is gained by humanity that advances us as a species and makes us better."

DOCTOR WHO:
LESSONS FROM A TIME LORD ON LOVING THE OTHER

Kate Muriel

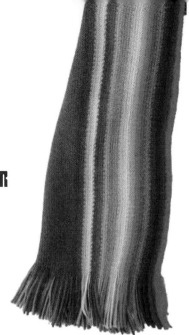

Every Doctor Who viewer has a favorite Doctor. With each incarnation of the titular character on television's longest-running science fiction programme, fans object to or rejoice in, criticise or celebrate the latest choice of actor playing the role of the ancient Time Lord. Though a canonical trait of the Doctor's race allows for a frequent change in casting and thereby for potentially innumerable stories to be told, the heart of the show is in fact this single character. The Doctor is so complex, so wonderful and relatable, that no matter the change in face and wardrobe, legions of Whovians continue to watch. One could argue that among the many reasons for its continued popularity, we are enamored in part by the way Doctor Who reflects back ourselves and the world around us from our screens.

After being off the air for nearly two decades, Doctor Who returned to television in a post-9/11, pre-7/7 world. The revival came in a time of escalating war in the Middle East, and as each successive season premiered, many came to the realisation again, or for the first time, of how science fiction can so closely parallel our lived circumstances. As Muslims became increasingly marginalised and immigrants gradually more abhorred across the Western world, so too did Who's fictional extraterrestrial invasions telegraph real-world anxieties about terrorism, open borders, and the perceived violation of a secure homeland.

Now in the age of border walls and Brexit, we continue to seek out fantastical entertainment, both as

a means of escapism and as a way of interpreting the strange socio-political space in which we find ourselves. We should question, however, whether the underlying themes of these types of media are truly making the impact they were seemingly meant to make. While Doctor Who is largely celebrated for its sharp and witty writing, clever character arcs, and beautiful plotting, it is also rife with allegory. Science fiction and fantasy have long written hard truths between the lines of their storytelling about the way humankind interacts with one another, and Doctor Who is no different.

Because many among us have made it clear that our spaces are only for those most like us, of course we cheer when the Doctor bravely returns an Earth-invading, non-human species to sender. We align ourselves with the Doctor, though she too is alien. After all, haven't we been taught to believe that the only good Other is the Other who makes themselves valuable to us at the cost of betraying their own kin? The Doctor's foes have, on occasion, been discovered to have once been her allies. Additionally, she has made countless interspecies and interracial friends, lovers, and companions. Yet it is when she stands first and foremost on the side of humankind that we celebrate her bravery the most. Whether consciously or not, this is because we subscribe to the notion of us versus them, the Ordinary People versus the Other.

In an effort to unlearn these biases, we must think critically about how we developed them to begin with, and we must charge ourselves to learn how to relate to those most unlike us. We mustn't just see ourselves as the heroic Doctor who gallantly saves a single planet over and over again. We must also see ourselves as the kind Doctor, the curious Doctor, the adventurous Doctor, the fearless Doctor. We are scared of the unknown; the Doctor charges headlong into danger. We think of our survival before that of 'those other people'; the Doctor asks us to consider that the 'invader' is often simply seeking refuge.

As flawed as any of us, the Doctor has at times given way to rage and needed talking down ('The Runaway Bride'), and has had to learn the hard way that others may misinterpret her actions through a much more fearful, human lens ('The Christmas Invasion'). Regardless, both the character and the show itself are lauded – rightfully so – for their ability to connect us, by way of an alien conduit, to one another as people.

Given this element of the show, it is not just the multi-faceted character of the Doctor we should aspire to be like. We must also find ourselves in the feared and hated Other. Across the world, the West has invaded other nations in the name of greed, war, and under the guise of 'liberating' those who have not asked us to do so. Uprooted by foreign occupation or simply seeking a different way of life, immigrants and refugees have made dangerous journeys, leaving everything they have ever known, to seek the fabled free world. When they arrive, they are greeted by masses of people who would rather see them sent back. To them, we become the Other, feared and reviled.

Ultimately, even if we only see the Doctor as the great protector of humanity, we need at least understand that the Doctor places no conditions on who falls under this protection. Her travelling companions have been women and men, black and white, queer and straight. They have been strangers and family, alien and human. Above all, they have all been among the stars and seen that there is space and time enough for all of us. Should we as viewers not acknowledge this as gospel truth?

We are one planet among many, one lifeform among so very many. We may never find a way to understand our animal friends or as-yet-undiscovered alien species, but if Doctor Who can teach us anything, it is that we share more commonalities with the very humans we name enemy than we can ever comprehend. With minimal effort, we can understand each other better than we ever knew we could. Unsurprisingly, love and compassion are the same in every language. We need only to open our spaces, and let them in.

KATE'S TOP 5 NUWHO EPISODES:

Series 1, Episode 7: 'Father's Day'

Series 2, Episode 1: 'New Earth'

Series 5, Episode 10: 'Vincent and the Doctor'

Series 10, Episode 2: 'Smile'

Series 11, Episode 3: 'Rosa'

FINDING HUMANITY
IN THE UNKNOWN

404 Ink interviews Brian Binnie, the first Scot in space

EVER SINCE HE WAS A CHILD,
Brian Binnie has been fascinated with flight, space and the wonders of the unknown. Fast forward forty odd years and a childish fantasy has become a very vivid reality, with Binnie making history through his piloting of the experimental aircraft, SpaceShipOne. He won the lucrative $10 million Ansari X Prize by reaching an altitude of 367,500 ft, setting a new spacecraft altitude record despite the numerous trials preceding the voyage. Perspective altered since his own space experience, Binnie's sights are now set on aiding the commercial space industry, connecting people to what once seemed so far from reach.

We discussed the ins and outs of his momentous journey, his plans for an upcoming book, and the tangibility of space tourism.

404 Ink: What first drew you to the idea of space exploration?

Brian: I blame that condition on my mother! From as early as I can remember, I've had a fascination with things that fly, and one day she asked what I wanted to be when I grew up. I was about 7 or so and after I fumbled around for an answer, she interrupted and said, "If I was a wee lad, I'd want to be an astronaut." Then she proceeded to tell me about space, planets, stars and rocket ships. Later, my parents gave me a picture of Neil Armstrong's first step elsewhere. So those seeds took root and all throughout my education and career, I always angled to put myself in a position to take advantage of an open door into that lofty arena.

How did it feel to become the spacecraft altitude record holder?

Well actually that record has an interesting story behind it. The night before my flight (a Sunday night), the Discovery Channel aired the first of two segments on the program documenting the flight. During what would later be commercial breaks in the program, the scene would switch to Burt Rutan's house where many of the SpaceShipOne team had gathered. Interviewing Burt was Miles O'Brien, CNN's space reporter at the time. He first asked Burt who was going to be the pilot and Burt wouldn't answer that question. Then he asked Burt how he thought it would go. And Burt without missing a beat declared, "Miles, not only are we going to hit a home run, but a Grand Slam!" (That's baseball talk for a home run with all the bases loaded.)

So, I'm at home pacing around, already nervous as a caged cat, and I'm thinking *Good God, isn't the bar already high enough without adding all this Grand Slam pressure?* Next morning, literally as I'm walking to the space ship, Burt pulls me aside and says, "I want you to swing smooth and go long!" While I'm trying to figure out what this might mean, he goes on to tell me that to beat the long-standing altitude record set back in 1963 by the X-15, I'd have to rise above 353,000 feet! Remember, the official bar for space at 100km required only reaching 328,000 ft so that was quite

"TO BEAT THE ALTITUDE RECORD SET IN 1963, I'D HAVE TO RISE ABOVE 353,000 FEET!"

a performance leap he placed on my shoulders – and it was the first I knew of there even being an altitude record out there to beat! If you go back and watch the videos online of that flight, you'll hear mission control advise me to shut the rocket down at a predicted apogee of 350,000 ft – which I dutifully acknowledged – but then took my sweet time to comply. As a result, the little space ship scooted well past the X-15's record, reaching 367,500 feet. There were three other world records set on that flight, but obviously the altitude one was the most significant. Now, in the US, it looks like the bar for space has been lowered back to 50 miles or 264,000 ft (no doubt, in part, to accommodate SpaceShipTwo's performance capabilities).

Were there times you anticipated failure or were you always confident in your success?

There was a period of 10 months after the first powered flight of SS1 that I flew, where it didn't look like I'd get a chance to get back into that cockpit. I was informed on a Thursday night that I'd be the pilot for the final flight for which we had completely revised the way in which we were going to fly the space ship. So, I was caught between the pressures of high performance and fine control (the previous flight departed controlled flight exiting the atmosphere – which would jeopardize Sir Richard Branson's interest in SS2), unless my flight demonstrated that we could reign in those demons. All these – changes, high expectations, my long absence from the cockpit and then Burt's final directive to go after the world altitude record – well, they were all tall orders that any number of issues or unknowns could have easily scuttled. Every one of the previous powered flights, five of them, had problems we hadn't anticipated, so there was really no reason to expect that the 6th one would

Family of Brian Binnie react as he launches into Space on the final SpaceShipOne flight
Credit: Don Ramey Logan, from Wikimedia Commons, CC-BY-SA 3.0

challenging because any number of things (beyond my control) could have ruined the flight – leaving only doubt and disappointment for all those gathered. Each of these challenges were further exacerbated by my personal trials leading up to the flight, which is a significant theme in my upcoming book.

Rewarding, meanwhile, was that it went off flawlessly, exceeded everyone's expectations of performance and precision control, and put an exclamation mark against the claim that even a privately funded, small team effort could accomplish what was previously the domain of governments and the Aerospace Primes.

be any exception, especially since we were introducing a brand new way to fly it.

What would you consider to be the most rewarding and challenging aspects of the flight?

Challenging was that some 30,000 people found the time and interest to travel to Mojave, on a Monday morning no less, increasing the town's population by an order of magnitude! Every news network was there with their satellite vans to beam to the world the events, however they might unfold. Cameras were everywhere, in the cockpit, in Mission Control and even the dressing room I used to get into my flight gear! Challenging too because just three days earlier we had completely changed the way in which we were going to fly the space ship into that magical realm. And let's face it,

Considering Richard Branson had an interest in SpaceShipTwo, do you think that space tourism will increase or detract from the mystery of space?

If done properly it will most definitely enhance the magical mystery of that realm. My only caveat would be that if you're in a confined cabin with five other people, all floating about in a haphazard manner or all vying for a view on one side of the vehicle, then the reality of dodging an elbow or knee from a fellow traveler could certainly diminish what is there to absorb. It is an experience that is deeply personal and as long as that is respected, the tourism industry should be self-sustaining simply through word of mouth. I feel very fortunate to have had the solitude to soak in that rich and life-changing experience.

"IT IS AN EXPERIENCE THAT IS DEEPLY PERSONAL"

What is the most beautiful thing you've seen in space?

The atmosphere. It is presented as an improbably thin, blue electric ribbon of light that separates the menacing black void above and the peaceful panorama below, 500 miles in any direction, that is home.

Outside of your career, what is your favorite space to relax in?

The golf course. Growing up in Scotland, I became an early devotee of the game. Fond memories still include summer holidays with my grandfather who would take me to St. Andrews. As I remember, a wee lad could play that course back then for half a crown!

You've mentioned your plans to put together a book on your SpaceShipOne experiences. Can you tell us a bit more about that?

A large part of a test pilot's training is to make understandable technical material for those without a technical background. While many books have been written about SpaceShipOne, none – in my humble opinion – have captured the fascinating landscape of the human trials, difficulties and drama that went hand in hand with the demands of flight testing a space ship. That's been my motivation since the outset and I believe I do it with humor, compassion, suspense and plenty of technical detail accessible to both the layman and the more technically astute reader.

On that note, if you could take just one book into space, what would it be and why?

That's a tough one. Assuming space in this case to be a destination and not just a day or two jaunt, then I think I would take the Bible. I'm no zealot, but if viewed as a workbook for how to behave and live, I think it could be a grounding reference as the human race attempts to venture off planet Earth.

Brian Binnie & SpaceShipOne

"BOOKS CAN CHANGE ENTIRE SOCIETIES WHEN USED RIGHT"

Imogen Stirling chats to Round Table Books' Khadija Osman

A 2017 research study revealed that just 1% of British children's books featured a BAME (Black, Asian, and minority ethnic) protagonist. The startling result prompted inclusive children's publisher Knights Of to launch a crowdfunding campaign with the view to opening a permanent inclusive bookstore in Brixton. The campaign was immensely successful, drawing donations from hundreds of supporters, and so Round Table Books was born. Taking the reins of the new venture is Khadija Osman, former bookseller at Forbidden Planet's London megastore.

404 Ink sat down with Khadija to chat about the importance of creating inclusive spaces within literature and whether we can be hopeful about the future of equal representation in publishing.

Khadija Osman

We jumped straight to the heart of the topic, asking Khadija to give her interpretation of the term 'inclusivity'. She defined it simply as, "breaking down the border than creates an 'us' and 'them' because we know that people are all of equal importance and that we all deserve equal respect." That being said, she acknowledged that this sentiment is simpler to preach than practice. "When we're not seeing everyone in a room together, being able to hold the same positions and command the same kind of attention, it's easy to forget that people are left out. Including those less able to speak up in our conversations means we'll remember to offer that respect and help where it's needed, in whatever form might come."

'Inclusivity' is a term frequently seeming to take the place of 'diversity', with the latter often deemed a buzzword that doesn't adequately serve its cause. While Khadija confirmed the personal limitations she finds in 'diversity', she is more interested in promoting the sentiment behind either chosen descriptor. "When I say we're working to ensure inclusivity in our bookshop I mean I want to include people of every previously excluded identity along with those who already hold the space and power into our shop together. I want the bookshop to work for anyone and everyone."

And what is it about a bookshop that holds the potential to set such a strong example of equality? "Literature is such an important medium to me," explained Khadija. "Art always has the potential to be amazingly powerful, but I think words just connect with people so well." Her passion for literature is evident as she stresses the importance of creating inclusive spaces within it. "The feelings conveyed in books can change entire societies when used right and if they're speaking about the experiences of races, identities and abilities then they can better the worlds those people are currently living in too."

Having previously read Khadija's comments that

"Round Table Books won't just be considered an 'inclusive' bookshop but simply...a bookshop", we were keen to learn her opinion on the tricky process of highlighting differences in order to reach normalisation – and whether this spotlighting is a necessary evil of the path to reaching a platform of equal representation. "I agree that it can build a real feeling of otherness when we start putting people's race or disability etc. before their career and achievements," she responded, "so it's something we've tried to avoid on the shop floor. With Round Table Books we want to highlight the love surrounding the titles that we do have and the normality of having them on the shelf.

"But while we have gone out of our way in the stock room to choose books depending on creative teams, themes and main cast, we're hoping that the customer who comes in asking for some teen horror or a board book is as helped and affected by the breadth of

"IT CAN BUILD A REAL FEELING OF OTHERNESS WHEN WE START PUTTING PEOPLE'S RACE OR DISABILITY ETC BEFORE THEIR CAREER AND ACHIEVEMENTS."

The work may be time-consuming but it's undoubtedly paying off, as Round Table Books has been met by strong public support right from its earliest crowd-funding days. Khadija put its public appeal down largely to "the fact that it involved children, the future and those with so little power over what is around them and the ideas put in front of them.

"A store that children can go to and see themselves in. Loiter in, if they're like me with bookshops when I was a kid. People were asking for something to change, not just for themselves but for all the little people in their lives. I love children's books myself and the changes I'm seeing in them, but it's when I think about the little ones around me getting to hold these stories in their hearts that I get all teary-eyed."

The opening of Round Table Books is clearly a significant move forwards in terms of the future of inclusive publishing, yet Khadija stated that while "we're making strides at the moment, a long way still needs to be travelled." She continued, "We see unfair and ignorant things happen every day in the world and in publishing and it's very disheartening. But I think with moves like that of Knights Of, to have an office in the heart of a bookshop and invite easy access to the masses, paired with more transparency and the will that I see in so many to enact change, then we'll be heading to a much brighter future indeed."

different characters as someone asking for a book with an Asian lead for their daughter to recognise.

"We're just trying to get you exactly what you want. Hopefully, in doing that we can come across as just a fun bookshop and not as a place making a spectacle of the remarkable talent on our shelves. Or at least not a bad spectacle. I am partial to confetti and the odd balloon."

"I AM PARTIAL TO CONFETTI AND THE ODD BALLOON."

With that in mind, I wondered whether the bookshop has found enough titles to fill its shelves, given the lack of inclusivity in publishing to date – and the bookshop's desire for a variety of genres. "We are trying to create a wide array of books from board, picture, first chapters all the way to young adult, and have them represent as many people and walks of life as they can," said Khadija. "So it's taking a lot of research but it's amazingly rewarding finding gems you can hold at the forefront of the store and stocking people who were – and are – making history with their works."

Round Table Books can be found at 97 Brixton Village, London, SW9 8PS.

Oor Big Braw Cosmos

A discussion with John C. Brown and Rab Wilson

Oor *Big Braw Cosmos* is hitting book-shelves with a (big) bang. A unique collaboration between Scotland's Astro-nomer Royal John C. Brown and eminent Scots poet Rab Wilson, the book fuses poetry, imagery and cosmic science to create a vivid insight into the work-ings of the universe. The wildly different backgrounds of the book's two creators, alongside their merging of seemingly unrelated subject matter, results in a book that is as fascinating in its own story as in its content.

The two men met approximately three years ago with their meeting being, as Rab describes, "serendip-itous". He goes on, "We'd both been invited to and involved in the opening of the Crawick Multiverse, a land art project down in Dumfriesshire." Rab had been commissioned to write a poem for the event which he presented to the group of visiting scientists; following this, John approached him with his proposal for a book fusing cosmos-based science with poetry.

The process of putting together *Oor Big Braw Cosmos* took around two years and combined independent work with frequent collaboration. "We spoke often on the phone, and by email and text," says Rab. "There was constant communication going on between us." John describes how they "wrote fairly independently but in stages, each itera-tively responding to the other's previous drafts." Rab continues, "It was like any partnership – you've got to have the same goal that stayed true all the way through the project."

Naturally, the pairing of astronomy with poetry demanded that each man deviate somewhat from his comfort zone. Yet, rather than pose a challenge, this seemed to further fuel the collaborative process. John found inspiration in "talking about our different worlds and experiences and trying to reach across that boundary in the subsequent round of writing." Rab, whose poetry had yet to venture much into the field of space, commented that "like most people, the idea of the cosmos, the universe...they're endlessly fascinating and interesting so it piqued my imagina-tion." Logistically, John would tend to feed Rab scien-tific knowledge and facts which then sparked Rab's imagination for the poems. "A lot of the poems came from John's research and writings in the book."

However, this role was sometimes reversed; John recounts one particular occasion when his own scientific knowledge was enhanced by Rab's poem 'Space Walk'. John says, "I didn't know anything about the Armstrongs as a Scottish clan, and Rab told me a funny story – there's a very famous picture taken by Neil Armstrong on the Moon and in fact it's the first ever selfie in space because you think it's a picture of Buzz Aldrin but actually Neil Armstrong is taking it and... you can see Armstrong himself reflected in Aldrin's glass helmet." Intrigued by the element of humanity introduced through Rab's poem, John adds, "I was motivated to write a longer piece about the Moon landings, stimulated by the poem." This toing and froing means of writing led to both men using the metaphor of jazz musicians playing off each other to indicate their mutual inspirations.

The book is undoubtedly novel in its genre. John stresses that "to sell it, publishers may want to know what it is primarily. But it's unique. Science and beautiful images interwoven with poetry. It's an experiment aiming to attract multiple audiences and to dissolve groundless cross-cultural reluctance." And the merging of forms seems be pretty seamless, with John making comments on how easy a pairing the two make. "All great scientific insights, especially in physics and astronomy, are beautiful and elegant as well as accurate and predictive – and well described by poetry." Rab continues, "You'd think it is something that should happen more often, this marriage between two different but connected disciplines." And it's evidently true – the book is strikingly beautiful in its finished format.

Rab proposes a thoughtful reason for why poetry works so well within the book and lends itself to the vast subject of space. "Poetry is a very special language. It's maybe the highest form of language, it's distilling something down to its essence. Casting a poetic eye onto something that is very oblique and strange and hard to grasp is sometimes the key to unlocking the difficulty in grasping something like that." With Rab's poetry being written almost entirely in Scots, there's an additional honesty to the language that resonates throughout the book.

As indicated by its title, much of *Oor Big Braw Cosmos* has a strong Scottish flavour, apparent both in its content and the language of the poetry. John comments that the Scots emphasis wasn't a slant that was hugely obvious during draft versions of the book but became a powerful focus in some chapters as they moved forward. It then reached the point that the men were "very much looking through a Scottish telescope", as Rab puts it. "There's quite a

Rab & John
(credit: Robin Gillanders & Glasgow University Photo Unit respectively)

lot of material in the book which relates specifically to Scotland, the ancient history of the country, its ancient tribes and standing stones." John describes that the book is set up to first provide a sound but accessible general astronomical overview before moving on to a Scottish perspective in order to demonstrate where some of the country's contributions fit in a wider context.

"The poems themselves, virtually all of them are written in Scots but there's a great Scottish sensibility about them, that Burnsian bawdiness at times. There's a rough and rude healthiness to them, a humour," Rab acknowledges. "Scots has this pithiness and punchiness about it that cuts through a lot of nonsense to get to the heart of something. You can say things in Scots that have got a warmth and a passion and a humanity about them."

Rab seems confident in the ability of poetry's

inherent clarity to convey the heart of the topic, a sentiment echoed by John when discussing his own speciality. "Some people ask what the use of astronomy is," he says, "and a reason I'm so fascinated by it is that in a way, apart from some spin-offs it's useless in a monetarist sense." He laughs in commenting that there is little real money-making in astronomy, though it has led some of the most immense scientific breakthroughs. Instead he sees its purpose as to observe beauty, to lift you out of the banal realities of the everyday and acknowledge something so much larger than yourself. The poesy of his statement is striking.

On asking both about whether it was challenging to delve into each other's fields, they each seemed to have tackled it well. John notes that while his interest in poetry is relatively recent, he's always been drawn to literature, music, art and magic and how they can be used to convey the beauty of science. Rab puts it simply that both poets and scientists are determined people, if there's a challenge to be met then they'll rise to it.

The conversation eased away from *Oor Big Braw Cosmos* to address space more generally. Both John and Rab recount their most amazing fact about space. John comments "how empty it is, though huge and packed with such beautiful complexity", while Rab adds the head-hurting fact that there are more atoms in your fingernail than there are stars in the universe. Asking each about their favoured non-work-related space to relax in, Rab describes cycling through the valleys around the Afton Water, a beloved space of Robert Burns, while John recalls the comfort found in relaxing to live jazz. They are finally met with the near impossible question for two literary-minded individuals: if you could take just one book into space, what would it be? John confidently answers *The Rubaiyat of Omar Khayyam* while Rab, true to the nature of the chat, offers: *Oor Big Braw Cosmos*.

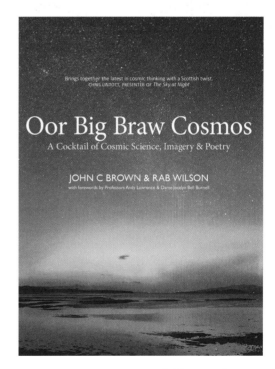

Oor Big Braw Cosmos is available now from Luath Press, bookstores and etailers.

BEHIND THE SCENES:
THE SPACE GECKO PROJECT

The Space Gecko Project is a unique collaboration of spoken word poetry, music and animated visuals, created by Stuart Kenny, Grant Robertson and Lewis Gillies respectively. An exploration of the love story between two geckos on a tragic space sex mission, the show is lauded for its heartening wholesomeness and was recently performed to a sold out audience at the Scottish Storytelling Centre.

So what is it that makes The Space Gecko Project such a standout show?
Stuart gives us an insight:

ACCESSIBILITY IS AT ITS HEART

"It's a silly show full of ridiculous puns, but I think puns can make a project really accessible. And it was written to be as accessible as possible, which is definitely something that the music and the art help with. They break down a lot of barriers for people."

EMBRACING ITS NICHE

"There aren't too many others doing wholesome animal-based multimedia spoken word shows! So we hope part of the enjoyment for people is seeing something new and a little different."

FUELLED BY COLLABORATION

"Grant and I live together and sometimes he'd come through with a guitar riff he liked, or a new instrument he wanted to use in a particular poem, and then I could write with that in mind. Other times I'd give a very vague prompt like "what if this one was a western" or "could this one please be a pop banger" and Grant would miraculously turn that into something not only listenable but very catchy."

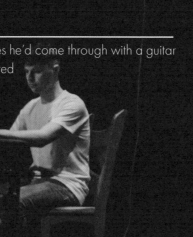

"THERE AREN'T MANY OTHERS DOING WHOLESOME ANIMAL-BASED MULTIMEDIA SPOKEN WORD SHOWS!"

EQUAL ARTISTIC WEIGHTING

"Obviously the words of the show make up the story, but the music really sets the tone and the sort of ambience of the show for the audience. You can give people a great idea of what kind of poem is coming up, and what to expect – whether it's a sad one, an upbeat adventure, or something more serious – just by a few musical notes. That's a lot harder to do just with words. So, we played around with how the music and the words paired, and what the show needed from each of them a lot. And then sometimes the performance or the delivery is tailored because we know Lewis has a particularly funny or heart-warming illustration that's about to pop up behind us."

CREATED BY FRIENDS

"We three have been looking for a way to collaborate for quite a while, as much just to have a passion project to work on together as friends as anything else, so this just slotted in perfectly, and it worked out really as well as we could've hoped."

The Space Gecko Project is now available on Spotify and iTunes.

SPACE IS COOL AS FUCK

Space is Cool as Fuck is an attention-grabbing coffee table book with a twist. A collaborative melting pot of artists and scientists, the book is co-authored by **Kate Howells** and serves to offer a brief overview of the mysteries of the universe – written in the filthiest language around. The book is visually magnificent, jam-packed full of artwork and photographs which complement Kate's own writing and the contributions from her numerous guest writers. It's a whirlwind tour of the wonders of space, with an enthusiasm that radiates from the page.

KATE HOWELLS' SELF-PROCLAIMED MISSION IN LIFE IS TO INVITE PEOPLE INTO THE JOYS OF SCIENCE

When we ask her to describe herself in a nutshell, she says, "I work for a non-profit called The Planetary Society that educates the public about space science and exploration and advocates for government funding for space. And whenever I can, I write about space and science for public audiences."

Space is Cool as Fuck is her debut book, which she summarises as "a wild ride through the universe, taking a quick look at a bunch of crazy cool phenomena that are mind-boggling but totally real." While tackling the universe may seem an immense topic for a first book, Kate instead stresses, "Honestly, it was a lot of fun. Writing is such a pleasure, and I loved the experience of exploring my own style. Plus, the research was fun to do because it was all on topics that I knew were really cool, so learning more about them was great."

Delving into Kate's initial inspirations, she comments, "I got hooked on space later in life than most of the other space nerds I hang out with. I was an undergrad, studying Psychology, and picked up a copy of Carl Sagan's *Cosmos* on a whim at a used book store. His descriptions of the two Voyager spacecrafts' missions to the outer planets totally blew my mind, and a passion was born." That passion was certainly realised as we now fast forward to see Kate working as Global Community Outreach Manager for The Planetary Society which she describes as, "a dream job." She goes on, "I get to meet fascinating people and learn about unbelievably cool topics, and working to share all of that with others is a real honour."

On that note, the book is unique in its collaborative vibe which sees a multitude of artists and writers lend their work to the pages. To begin with, it makes the book absolutely stunning. "We really wanted to convey the crazy, awesome, beautiful nature of the cosmos, so what better way than with crazy, awesome, beautiful imagery?"

Kate Howells (probably swearing)

Notable in the written contributors, meanwhile, is the variety of sources they come from. Experts and others collide to create an engaging and wide-reaching sense of collectiveness. "With the guest writers, I really wanted to convey that anybody with an interest can learn about science and share their knowledge with others - you don't have to be an expert," says Kate. "And bringing so many talented artists on board was just such a pleasure. I'm lucky to be friends with a lot of really brilliant artists, and it was great to be able to work with some of them on this project."

One of the book's collaborators is the one and only Bill Nye, whose lengthy and insightful interview is a pleasure to read through. "It was such a treat," Kate beams. "I work with Bill and have been able to have a lot of really great chats with him, but I'd never sat him down and picked his brain like this. And he was such a great sport, giving us so much of his time and really having fun with it. He's got a great sense of humour and a wonderful perspective on things, so getting to explore that was fantastic. Plus, we got a childhood hero to swear, and that's always fun."

Ah, the swearing.

WE GOT A CHILDHOOD HERO TO SWEAR... THAT'S ALWAYS FUN.

Space is Cool as Fuck (as the title may suggest) uses bawdy, colloquial language throughout. When asked the reason for this, Kate answers, "Reading this book was meant to come as close as possible to the experience of sitting next to me at a party, a couple drinks and maybe a joint or two deep, and listening to me rave about the latest cool thing I learned about space. So, the language I used was the language I'd use in real life – sweary, slangy, jokey. When something is fucking cool, you have to express that. The added bonus, of course, is that by using unusual and potentially alarming language you grab people's attention, so it's all part of the plan to trick people into learning about science." Her writing imbues the book with a vibrancy and humour that allows an accessibility and a breaking down of academic barriers. This is a recurring theme with Kate who seems determined to normalise space, retaining its mystery

while heightening its approachability. She claims that the book is written "to reach people who don't normally pick up science books", explaining that, "it's so common for people to think that science is going to be boring, or difficult to understand. We wanted to announce on the very front cover that this wasn't going to be the case." The book is designed with the very purpose of standing out from other space-themed work, with "the humour, the artwork, and the incessant reminders throughout the book that you – yes, you – can be a science person if you feel like it."

The sense of fun that *Space is Cool as Fuck* instils is infectious. Asked if writing one particular part of the book was most enjoyable, Kate replies, "I think my favourite chapter to write was the one on dark matter and dark energy. I really had to work hard to wrap my head around these concepts enough to be able to explain them in a clear way, and that was a fun process. I can only hope it makes sense to readers!" We were curious to ask Kate for her most mind-blowing space fact. "Some of my favourite facts about space are actually about the Earth as a planet. We tend to think of space as being all 'out there,' so when we're reminded that we're a part of it, that can be extra mind-blowing. So here's a good one: every single day, the Earth gets bigger. As we move through space in our orbit around the Sun, we encounter random bits of space rock that hit our atmosphere, burn up, and settle on the planet. These space rocks are mostly pretty tiny, but there are a whole lot of them – we grow by 100 tons every single day."

And if readers are to take one thing from Kate's book? "That science isn't dull, and that space is cool as fuck!"

Space is Cool as Fuck is published by Lost the Plot and is available now.

THE DAY OPPORTUNITY DIED

A perspective on the popular Mars rover's end by Errol Rivera

Image: NASA/JPL-Caltech/Cornell/Arizona State Univ.

One day, Opportunity, the Mars Exploration Rover died and I hated the internet.

Then I thought *NASA is art and we are all space robots.* For writers, it can be very tempting to write crap like that. You have an experience, do the emotional math in your head, put it into a clever phrase and call it a day. Thankfully, when I read back the Opportunity's last update, "my battery is low and it's getting dark", I'm reminded that the math is not the message, and making a connection with people is more important than being clever.

If you were internet-ing not long after NASA shared the final words of Opportunity, you likely saw an overwhelming torrent of love and affection that swept across the face of Twitter. Or maybe it was just my feed – it's Twitter so you never can be sure. But nevertheless, I wasn't the only one blown away by how much people loved the little guy. In fact, techradar.com wrote a lovely article asking that very question – how come we loved that robot so much?

The cynic in me has a dark answer to this question.

How many times in life have you felt like you're the only person to really appreciate a public figure, forging the rarest of connections with people who actually understood, or espousing their brilliance to blank faces at a party? This could easily go on your whole life, until one day that public figure dies and suddenly you're forced to witness those same blank faces fill your social media with overused press photos of your hero and long testimonials about how inspirational they were (bonus points for length and use of rarest photograph). Maybe Opportunity, literally the toughest little rover in this world, was only loved because it died.

I can distinctly remember seeing one of those adorable Opportunity cartoons tweeted by someone I can't stand and looking through their follow list like, "do you even follow NASA, bro?" They totally didn't. I wouldn't have been so annoyed if they hadn't used one of those really sad cartoon illustrations. The one with Opportunity's last words. I hate how sad those last words make me. I'm pretty sure those words actually weren't about batteries and darkness but probably a string of numbers, graphs, and status updates from its various systems. So how come I and so many others were moved by those stupid cartoons and those stupid made up last words? Because they weren't made up. They were imbued.

And that's when I was forced to eat my own darkness. The truth is, I'm not very cynical at all, just like NASA isn't just numbers and graphs and systems. Yes, Opportunity was just a robot that shut down. But not to me. Not to a lot of us. And certainly not to the people that built it, talked to it, listened to it, and were surprised by it when it surpassed its mission parameters of 90 days and operated for 15 long years. When the Opportunity team received the last set of numbers, and graphs, and system status updates in June 2018, it was more than information from a robot. They were reading a letter written in their own language. A letter from a friend who was tired and cold, and alone. After years of being part of that little rover's story, the Opportunity team wanted to share what those last words meant – the kind of meaning only life can imbue. That's what happens when you understand someone, even if they're made of different stuff and live a world away. They come to life because you speak their language and hear their stories. Opportunity wasn't a robot that shut down, and it wasn't loved because it died. It died because it was loved.

To me, NASA has always been a great story about passion, effort, desire, and creativity, because I'm a space nerd for sure. Right now, my favourite thing in the world is an app that lets you do rocket science without having to know math. I landed a robot on Mars last week. His name is Lars Grover, the Mars Rover, and he's dope.

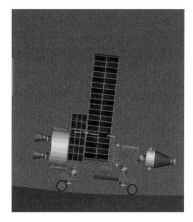

"Lars Grover, the Mars Rover"

It's the kind of thing I would have died for when I was twelve and building my NASA shrine. All of my clothes had to be kept in a rickety unvarnished dresser, even the nice shirts that were meant to be hung up, because I needed the space for my actual shrine. That's where I kept my copy of Carl Segan's *Cosmos*, my VHS of Apollo 13, and my replicas of the Space Shuttle, the CM, the LM, and the EVA suit. It's also where I kept my genuine NASA Space Shuttle handbook where I learned all those acronyms (command module, lunar module, and extra vehicular activity, respectively). Any space nerd is going to take a disproportionate amount of joy out of learning all those terms. If Opportunity was around then, I would have had that too. I probably would have printed out a picture of the path it took in its unexpectedly long life. I would have taken a red marker and circled the crater where it shut down and written the name in big red letters – Perseverance Valley. I probably would have looked at it and sighed. Then I would have memorised every Opportunity-related acronym I could and explained them all in detail to my mother. I know this for a fact, because I did that with every other major space vehicle (sorry, Ma). As a kid, I would roll my eyes at car-guys talking cars, and sports-guys talking sports, then unironically turn around and lament to my mother about the shelving of the revolutionary Aerospike engine and how it would have replaced the Space Shuttle's fuel rocket system. All totally blind to the fact that anything meaningful I had to say was disappearing under a mountain of jargon (its memories like these that make me worry all men are truly the same).

It wasn't until I was older and started writing that I realised what I really loved about NASA and space travel was the stories. In fact, I'd wager that most creative people love space for the same reason. I'm talking about telling stories, calling up powerful imagery, giving shape and name to things that were once only feelings. The kind of creativity that drives you to evoke a unique experience when nothing else will do. Why else would the first mission where two vehicles docked with each other in orbit, be called 'Gemini' – the god of duality and wholeness? And why else would you take numbers and system statuses and turn it into a tear-jerking goodbye from a lonely little rover? Why would you do that unless you experienced something so meaningful that you had to share that with others? I thought that as a kid, my love for NASA made me want to speak their language, when the reality was that I loved NASA because they were trying to speak ours.

On February 13th 2019, the Opportunity mission was declared complete. All of its data was collected and then some, but the people who experienced it and were changed by it wanted us to understand what it meant to them. They told us a story about their friend, an explorer they talked to every day, who went longer and further, and learned more than anyone expected, before it finally rested in Perseverance Valley. That's something we can all connect with. It means something to me, to you, to the writers in the magazine, and definitely to the women who published it. As metaphors go, it's almost a little too easy. That's probably another reason artists love space so much, especially writers. Few things are as inspiring as a story that tell us to keep roaming and to surpass our mission parameters, but there's something more to be learned here. Not from the story, but from the telling.

Yes, go on your adventures, send back messages, and keep sending them, but more than anything remember to let people in on your language and try to speak theirs. One day, after you've laid down in your own valley of perseverance, we can all trace your path, and even though you spoke in your own unique way, we'll hear your words for what they meant. The way they were heard by the people who loved you.

RIP Opportunity. Image: NASA/JPL-Caltech/MSSS

the 404 Ink corner

Welcome to the 404 Ink corner.

We previously kept this magazine as its own separate outlet from our books, but we felt it fitting to bring it all together into a wee corner to keep you updated on the latest in 404 news and author updates in each issue. What can you expect? Well, continue on and find out...

THE HIGHLIGHTS

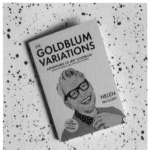

BBC Scotland have adapted three stories from Chris McQueer's debut *Hings*, published by ourselves in 2017. They're amazing (though not for the faint-hearted or easily-queasy) and available on the BBC iPlayer. Chris' books will also be coming as audiobooks to Audible soon.

Penguin Books will be publishing Helen McClory's *The Goldblum Variations* in America while we'll be releasing a revised edition with brand new stories here in the UK at the same time. Since we last spoke, Jeff Goldblum actually read stories from the book in Germany. No big deal.

Helen McClory's *On The Edges of Vision* was picked by Ali Smith as one of her reads of the year – 'a writer completely unafraid', and Nadine Aisha Jassat's *Let Me Tell You This* was picked by Jackie Kay – 'a powerful debut.' Nice.

Coming September: explore the issue of human rights through getting to know the inhabitants of one street – *Constitution Street*. An important and timely book from human rights activist Jemma Neville. Available at 404ink.com.

We had our biggest launches to date. Over 120 people showed up to launch *We Were Always Here* and 170 for Nadine's debut poetry collection, both with our favourite radical bookshop Lighthouse Books.

We have a new book! Euan Lownie's *Never, Ever Take Anybody's Advice on Anything* publishes in November featuring advice for life and careers from many at the top of their respective careers in Scotland.

A catch up with Chris McQueer

Photo: Sinead Grainger

How has 2019 been for you so far?

2019 has been magic so far. It kicked off with writing the final drafts of the *Hings* TV scripts, editing them, then right into filming which was hectic and chaotic but the best fun ever.

How did you find the process of transferring *Hings* from page to screen?

It was really challenging. Obviously in a short story on the page, you can make anything happen. As long as you can describe things there are literally no constraints, whereas adapting stories for the telly, there's a lot of stuff that just can't be filmed or would cost a fortune and take forever to film. It was frustrating having to cut stuff out, but it was a good challenge and writing in a totally new medium was great fun.

Did you have much involvement during the filming?

Aye, I was on set for the three shoots. I got to listen in and sort of shadow the directors Kris and Dave and watch and learn from them. Sitting with the actors and hearing how they planned on delivering lines was interesting as well, hearing how they used their accents and intonations in their voices to comic effect.

Did the stories turn out as you'd visualised them or was there an element of reinterpretation?

With these stories I wanted to try and make them better and funnier than they were in the book. I wrote the original drafts of these three stories maybe two and a half years ago and my writing has changed a bit since then, so I took adapting them for the telly as a challenge to see if I'd improved. So, I re-wrote a lot of the dialogue and played about with the structure of the stories to try and make them better. I think straight adaptations of books to TV or film can be boring and I wanted these to really capture people's imaginations, so letting the directors, cinematographer and the actors have a lot of input and hear their interpretations was really important to me.

How did you choose which of the stories to focus on?

These were really the easiest ones from the book to film and also the stories that could be told within the short run time we were working to. I felt they were good stories to bring people into the world of *Hings* and show them what I'm all about.

What has the response been like to the shows?

It's been good so far, the reactions from people on social media have been so nice and I'm glad people are taking time out of their days to watch them and appreciate the hard work that went into them. Hopefully we'll get to make some more.

Any recent TV/books that you'd recommend?

Nadine's collection, *Let Me Tell You This*, was sensational. I watched Russian Doll recently and I think it had the best ending of any TV show in recent years. Also, if you can, definitely go and see Beats as soon as possible, what a film.

The three stories from *Hings* are available now on BBC iPlayer.

Photo: Chris Scott

The power of heart and voice with Nadine Aisha Jassat

Nadine Aisha Jassat is a poetry revelation. 404 Ink first worked with Nadine through our debut book Nasty Women, and we had always said: "If she ever does a poetry collection, we have to publish it." And so we did.

Let Me Tell You This published at the beginning of 2019 and is an electrifying collection. Jackie Kay called it 'a punchy, powerful debut', Hollie McNish 'really likes Nadine's poetry', and Sabeena Akhtar says that 'if you read one poetry collection this year, let it be this'. Nadine takes readers on a journey exploring heritage, connection and speaking out, showing the power of heart and voice in a beautiful book that sits with you long after you turn the last page. If you're new to Nadine and her work, here's a quickfire introduction to one of the most brilliant and exciting poets in the UK today.

The journey.
"I think I've always been a writer, since I was a child crafting my own stories or stood on a chair in my parent's house at nine years old reciting poems. In my wider life, I've believed time and again that there's a relationship between voice and personal healing, and wider still in storytelling and political change. *Let Me Tell You This* was a journey through all of these selves; the writer, the activist, the storyteller, the person."

On her poetry's themes.
"Broadly speaking, my collection covers themes of heritage, childhood, connection, racism, women's stories and voices, and gender-based violence. However, ultimately, it is about speaking out and telling your story, about the power and freedom that comes from that. These themes are all part of that story for me, but it was the story and self-expression ultimately that mattered above all else. That it resonates with and helps others is something I find deeply touching, and fills me with awe."

Why poetry?
"I write across poetry and prose, and perhaps broadly I am a poet and a life writer; I write from my life and my experiences, and treasure playing with voice and telling stories in poetic ways. I find that some of those stories need to be told in poetry, some in an essay or narrative non-fiction, and some in short stories. What I like about poetry is that in each poem you can concentrate and distill your focus into one moment or aspect of something, and give it its own space."

On the page vs. on the stage.

"Often, readers will say that reading it aloud makes it 'come alive' more and adds additional layers of meaning or of emphasis. That said, I do enjoy playing with form on the page; for example in *Let Me Tell You This* the poems 'Threads' and 'Hopscotch' both use form very deliberately to explore and emphasise what each poem is about.

"In the moment, when I first write a piece, it's about exploring/capturing/processing or recounting a particular image, emotion, thought or moment in time. It's very much just me and the page. The way I write, however, is

Nadine reading at her Edinburgh launch (Photo: Chris Scott)

instinctively with a rhythm – it guides the piece without me even thinking of it – even my free writing goes this way, and leans toward a love of sound. This is something that I've only just begun to learn or realise about myself, even though I've been doing it all along; its very exciting to think of writing and my relationship with poetry as a journey or an adventure, as something where I keep learning more about myself and my work each day.

"After getting the words down, I'll then read a poem out loud to further find its rhythm and make sure the words meet it. If I've been writing the poem in my head before putting it on the page, it might already have a smooth spoken rhythm, but when the rhythm doesn't flow in my voice I find it as jarring as if I'd used a word which doesn't fit what I'm trying to describe. When I then read the poem to an audience, I think of it as telling a story in all of its components: the rhythm of the piece, the words, the emotion, how I tell it. It's all part of telling the story of that poem."

The inspiration.

"In my early days, before I could even imagine being where I am now as a writer, watching Andrea Gibson on YouTube helped bring me back to poetry and helped me see poetry in a new way. Their work has so much power and emotional resonance, and is so intelligent and beautiful, too! It made me realise that there was space for my voice, and for my work, within which the power of emotion and the spoken rhythm of the piece is just as important to me as the imagery and language. In the UK, I've long admired Hollie McNish's use of language and the way her poetry opens doors and breaks down barriers about what poetry is and who it is for."

Let Me Tell You This is available from 404ink.com.

We Were Always Here: still here

An interview with Michael Lee Richardson

We *Were Always Here: A Queer Words Anthology* has been out for about six months at the time of writing, and what a time it's had. A branch of Queer Words Project Scotland (led by Michael Lee Richardson and Ryan Vance), the anthology has been met with incredible success, with packed launch events and a reprint now in progress. Lauded for its showcase of queer writing talent, *We Were Always Here* is a timely, poignant and entertaining collection.

404 Ink chatted with co-editor Michael about the book's creation, reception, and future.

"It was incredible," Michael comments on how the book came into fruition. "When you put out a call for submissions for a project like this one, there's always part of you that worries you won't get enough submissions – or enough good submissions! – to make a book. And as the deadline drew closer, our inbox was still pretty sparse, so it was a tense couple of days for a couple of worriers like myself and Ryan! But, true to form, a lot of people seemed to have left it until the last minute - we even had a healthy handful of submissions sneak in after the deadline, so the theme of queer time and temporality extends to more than just the content of the anthology!"

Fears of insufficient material quickly dissipated; the editors were left instead with an array of stellar work. "We had a pair of real standout pieces early on, and there was a point when Ryan and I were both DMing each other – have you seen this one, have you got to this other one yet – which was really exciting. Reading stuff like Heather Valentine's 'Projector', Ciara Maguire's 'The Middle of Everything' and Andrés Nicolás Ordorica's 'IKEA' for the first time was such a joy."

With its striking bright pink leopard print cover, *We Were Always Here* gives the impression of vibrant positivity, yet Michael explains this wasn't the direction the book ended up taking. "Actually," he says, "a lot of the submissions were 'weird' or dark, in their way, and it's something that readers have really responded to. It feels like when we're given scope to write about our own lives and experiences without having to go over some of the Queerness 101 stuff, there's the scope and the space to really talk about stuff that means something to us, as a community."

"I keep saying that the big Scottish queer anthologies say something about the time they were published," he continues. "*And Thus I Shall Freely Sing*, from the end of the '80s, during the AIDs crisis, has a focus on sex and sexuality; *Out There*, from 2014, when the Equal Marriage campaign was really coming to a pique, has a focus on themes of love and family and relationships – and I'm interested to see what people think *We Were Always Here* says about 2019, when we look back in five or ten years' time."

Of course, Queer Words Project Scotland existed before the anthology's publication, initially in the form of a mentorship programme for emerging queer writers. Expanding on this, Michael says, "Ryan and I had both talked about how, as queer writers, we're sometimes – not always, but often – constrained by the narrow scope in which straight 'gatekeepers' will understand our stories or allow our stories to be understood. That can often mean that the stories that are allowed to be seen or heard are the ones that can be understood from a straight perspective. With writer development programmes and mentoring, even with editors, sometimes there's a real lack of understanding of our lives and culture, so even those writers that get through the gatekeepers are on the back foot. The mentoring programme paired 5 emerging writers with five established writers, and each pair got a couple of mentoring sessions to workshop a piece together and discuss career development."

The programme flourished, as did the writers, with Michael using emerging poet April Hill as an example of its impact. Through guided mentorship by Rachel Plummer, April moved swiftly from a first open mic performance to a headline poetry spot, a real triumph for such a new writer.

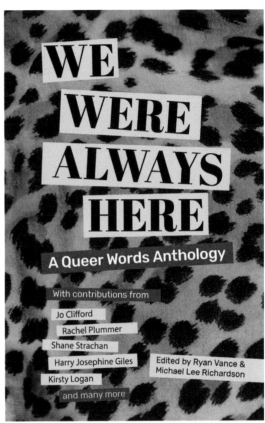

Similarly, Michael comments on the blossoming careers of many of the *We Were Always Here* writers, as he reflects on the anthology six months on. "When the book was first published, I said that I thought we'd look back on it and see so many people who had gone on to bigger and better things, that it felt like it might be a bit of a time capsule for the Scottish literary scene. I don't want to suggest that I'm an extremely powerful soothsayer or anything (I am, but I don't like to brag), but we can see that already start to bear out, even in the six months since the book was published. Alice Tarbuck is one of this year's Scottish Book Trust New Writer's Award winners, Christina Neuwirth's novella *Amphibian* is doing incredibly well, Garry Mac - who wrote our foreword - had a piece featured in GoMA's Queer Time School project that's now going to be in the GoMA archive."

After a period of such solid work, the editors are taking a well-earned rest – yet this is by no means the end of the project's journey. "We've got some interesting events coming up under the *We Were Always Here* umbrella," Michael says. "I'm keen to do more stuff that feels engaging and accessible – I really want people to come to our events and have a nice time! I'm also keen that we continue to support individual writers, especially trans writers. I've also been chatting with a few people about something that brings this work together – a festival or something – but that's at the very, very, very early stages."

On that note of rest and downtime, I ask Michael for any current book and television recommendations. He answers, "My focus has been on reading more trans writers this year. I loved *Small Beauty* by jia qing wilson-yang, a sharp and witty Canadian novel about family and grief and the small things that make up a life. The characters - especially the protagonist Mei and her friend, Annette – are so well-drawn and keenly-observed. Another book I've enjoyed this year is Hal Shrieve's *Out of Salem*, a witty '90s YA featuring a lesbian werewolf and a non-binary zombie witch that's definitely for fans of Buffy and Harry Potter. On telly, Years and Years, Russell T Davies' new show that takes place five, ten, fifteen years in the future, is really worth watching. And anyone who still hasn't watched Pose should definitely set aside a couple of evenings to work their way through the first series before the new series starts later this year."

We Were Always Here is available now from 404ink.com.

The messiness of reality: Elle Nash on *Animals Eat Each Other*

What does it feel like to lose yourself so completely to someone else? What does it look like to do so for the wrong people? It's these threads of obsession that flow through Elle Nash's Animals Eat Each Other. We follow a young woman with no name as she embarks on a fraught three-way relationship with Matt, a tattoo artist, and his girlfriend Frances, a new mum. Dubbed 'a cartography of eros, a heart-bomb' by Lidia Yuknavitch, her debut has been drawing attention for its unabashed honesty in the brutality of finding yourself, and that all began as the seed of a short story.

"I started by wrapping a narrative around a particular painful feeling – heartbreak, feeling lost, unsure, and stuck in living the expectations everyone else had for me – and expanded around that feeling," recalls Elle. "At the time I had been working with my mentor Tom Spanbauer for a bit. We would meet via Skype every two weeks to discuss the story. Eventually it just kept expanding and expanding until it became book-length. I wasn't even sure when I had committed to writing it as a novel, it just sort of grew."

"I had written a series of poems in an earlier chapbook, one in which one of the lines was 'animals eating each other'. I made a video poem to go along with the release of the chapbook in which I whispered this line over and over again behind the poem itself, and just felt haunted by it. To me it was a simple fact of life but one we ignore, as humans, who like to pretend we are above the fray of the natural cycle."

> # I WANTED TO MAKE A NOVEL THAT FUNCTIONED AS A MIRROR

Though technically nameless, so far does she lose her identity, Elle's protagonist is dubbed 'Lilith' by those she desires the most, a name she fully embraces. "Lilith was the first wife of Adam," explains Elle. It is not a story that is told in the Christian mythos – she's usually overlooked, but in Jewish folklore – she even appears as far back as ancient Mesopotamian folklore – she was made from the same clay as Adam rather than pulled from his body the way Eve was. This is such a significant metaphor, because the story goes that she refused to bow down to Adam's authority. Because of this she was cast out of the garden.

"Throughout ancient history Lilith was portrayed as a demon who represents wanton sexual desire and was rumored to have 'stolen newborn babies'. It is interesting to me, that a historical female myth refused to submit to authority, and is then assigned the caste of demon. I think that was one of many reasons I wanted the main character to be named this – she's given this second class caste in a way, almost as if she herself has been cast out of an allegorical garden, which is the acceptance and love of the couple, Frances and Matt."

That desire to be accepted is one that readers can easily recognise. "I wanted to make a novel that functioned as a mirror," she says. "I've had people tell me they could relate to almost everyone in the book – even though many of them do unlikeable or manipulative or hypocritical things – but wanted to shine a mirror on the sort of behavior we're all capable of as humans. We all experience longing and alienation, which is fascinating because it often feels so lonely and singular.

Animals Eat Each Other stands apart as it is messy and often without answers; the darker elements at the core of the book surrounding loss of resolution and loss of self are incredibly perceptive. "I wanted to create something that was realistic. One of my fundamental beliefs is that closure is a myth. It's something that one has to make for one's self. I wanted this to be reflected in the book – there are many issues Lilith has, for example, but none of them are the core of her identity. She simply has these facets (drug use, her sexuality, manipulation, feelings of abandonment, her sense of being directionless) that all come together to make her identity. Growing up I abhorred the sort of Lifetime-esque movies of teenage difficulties that focused on singular issues, like drug use or coming out as gay or bisexual, or eating disorders, as if these were the only functions of a teenager's identity, and that gaining acceptance or recovery or something somehow 'fixed' these errors in their personality. Reality is messier than this, and much more complicated."

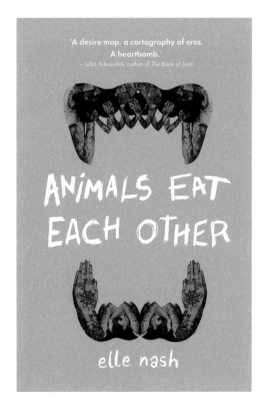

That messiness of reality filters through until the end, in a sense. "Sometimes people just don't change, they continue on with their bad habits, even after learning specific lessons about said habits. When I finally got to that point in the story, where I'd typed the words out, I sort of took a step back to look at it, and it felt right to me."

The book has been out for over a year in America, where it was first published, and came to the UK in Spring 2019 from 404 Ink. How has the journey been? "The most enjoyable part was sharing the experience of it with other readers. I'm so glad that even one person found the book to be relatable or otherwise helped them think about relationships and humanity."

"Once it was out in the world, I knew I couldn't control it anymore, or control how people reacted to it. I kind of just wanted to set Lilith free. I think, like with most things, I hope people take whatever they need to from her, and from the novel itself."

Animals Eat Each Other is available now from 404ink.com.

The best 'space' books around...

Everyone's a Aliebn When Ur a Aliebn Too – *Jomny Sun*
Jomny is a lonely alien sent to study Earth. Here are his illustrated adventures as he meets a mix of creatures and gets to know their views on life, love and happiness. It's so adorable and calming and pure.

The Comet Seekers – *Helen Sedgwick*
A lifetime feels like a long time, but it's barely a blink of an eye to a comet. Dipping through the past of the characters, and their families, only when a comet is in the sky, The Comet Seekers manages to encompass a thousand years in a few hundred pages and have you involved in the characters every step of the way. Magnificent book.

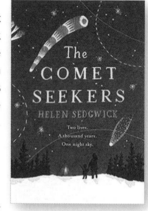

An Astronaut's Guide to Life on Earth – *Chris Hadfield*
Through his childhood ambitions to be an astronaut and refusal to give up, through extensive training and living for months on the International Space Station, this is packed with loads of little anecdotes that raise a smile. Whether Hadfield's chatting about his infamous Bowie video, or the nitty gritty of being an astronaut, the book acts as a reminder that space is cool.

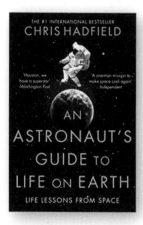

Space Is Cool as Fuck – *Kate Howells*
We're all here because we think space is cool as fuck, so why not learn more about how bloody great space is in a sweary but infor-mative book? Sounds fucking fantastic, and we can confirm: it is.

The Book of Joan – *Lidia Yuknavitch*
Joan of Arc and Christine de Pizan are cata-pulted to a near-future in a tale of revolution, in which humanity is being rewritten through propaganda, lies, and experimentation, all while the earth below lays crumbling. It's a tale of our times and a whirlwind of a book.

Chasing the Stars – *Malorie Blackman*
Olivia and her twin are heading back to Earth after the rest of their crew are wiped out; Nathan is part of a community heading in the opposite direction. Their lives collide, and so their journeys and plans change. It's a retelling of Othello, set in space, and it's really rather good.

...and these are pretty good too

Sure, it's all good and well talking about space books, but what is a literary magazine without a space to talk about books at large. Here's a handful of fantastic books we've read lately that we think you'll like too.

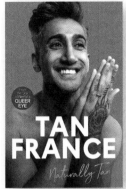

Naturally Tan – Tan France
Tan France says that his book "is meant to spread joy, personal acceptance, and most of all understanding. Each of us is living our own private journey, and the more we know about each other, the healthier and happier the world will be." That sums it up. A really wonderful book.

On The Come Up – Angie Thomas
Sixteen year-old Bri wants to be one of the greatest rappers of all time, or at least make it out of her neighborhood one day. Angie Thomas is back with her razor sharp writing; an incredible story of fighting for your dreams when the world will stand in your way. A homage to hip-hop, and another home run from Angie Thomas.

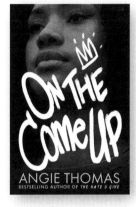

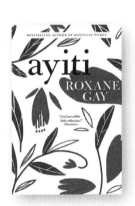

Ayiti – Roxane Gay
Roxane Gay's debut collection has been re-released to the world, showcasing the Haitian diaspora experience. Her short stories pack a powerful impact, the longer ones sweep you along, with pages flipping in quick succession. It seems Roxane just crushes it in every form. Who knew? (We did, but still.)

To All The Boys I've Loved Before – Jenny Han
The movie basically took over the world when it went on Netflix. Going back to read the books, it's pretty much the perfect adaptation, following Lara Jean as all the letters she'd secretly written to her great loves and crushes are unleashed into the world. A fun read (and fun movie).

My Sister, the Serial Killer - Oyinkan Braithwaite
Korede's dinner is interrupted one night by a distress call from her sister Ayoola - she knows just what to do. You'll likely never read someone be so blasé about being a serial killer, but it's brilliant. Precise writing, little fluff, matter of fact death, life going on. Disturbing and witty.

How To Come Alive Again – Beth McColl
Beth McColl's guide to killing your monsters is a mental health guide of sorts that never feels like it's telling you how to live your life. It speaks from experience, offers multiple solutions, acknowledges the good and the bad. Never a quick fix, instead it's willing to walk the long journey with the readers who need it most.

Subscribe to the 404 Ink Magazine by joining us on Patreon
patreon.com/404ink
There you'll get the magazine first, before anyone else or any shops!
You'll also get some behind the scenes content, be part of a wee community
of like-minded readers, and more. Because you're generous and we like you.
You're keeping this magazine alive and we can't thank you enough.

You can also buy individual copies of this magazine or our other books
directly from us at **404ink.com/shop**

That's all from us. See you in December 2019 for issue 6 - EARTH!

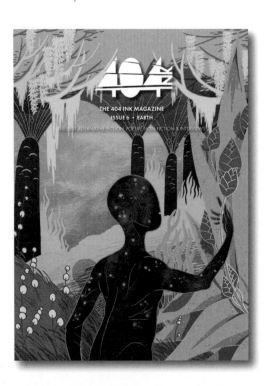

**With thanks to our Patrons,
old and new:**

Alistair Braidwood
Amanda Palmer
Aran Ward Sell
Ashley Wyse
Caroline Clarke
Catriona Cox
Chris Boyland
Claire Genevieve
Claire Squires
Claire Withers
D Franklin
Dean M Garland
Elizabeth Stanley
Ellie Otomiya
Emma Zetterstrom
Finbarr Farragher
Gordon Hayden
Hew
Iain Hoey
Iain Ross
Inside The Bell Jar
Jamie Norman
Jean Gray
Jean Marie
John Dexter
Julia Koenig
Kerry McShane
Kirstyn Smith
Kris Haddow
Kristin Walter
Lyndsey Garrett
Madi
Mairi McKay
Nicola Balkind
Nicole Brandon
Noel Johnson
Paul M. Feeney
Paula S Carr
Peter Kerr
Rod Griffiths
Rosie Howie
Sean Cleaver
Simon Brown
Simon Rowberry
Stephen Paton
Stevie Williams
Susan McIvor
Suzanne Connor

Goodbye.